Siem Haffmans – Marjolein van Gelder – Ed van Hinte – Yvo Zijlstra

BIS PUBLISHERS

ACKNOWLEDGEMENTS

The inspiring book *Products that Last*, published in 2014, is used frequently for workshops and projects at Delft University of Technology and for the CIRCO organisation. *Products that Flow* is meant to be an addition. During our workshops and classes for companies, designers and students, we discovered that we were lacking a framework for fast-moving consumer goods. Therefore, we decided to write the book *Products that Flow*. We are grateful for the supportive cooperation of the authors of *Products that Last*: Conny Bakker and Marcel de Hollander.

PRODUCTS

The authors would like to thank all the team members and trainers of the Dutch CIRCO project, *Creating Business Through Circular Design*. This successful CIRCO project has already reached more than 300 companies with a three-day workshop, and more than 250 designers with the CIRCO classes. All the participants of these CIRCO workshops are also thanked for their enthusiastic participation and input, that helped us to create the framework of *Products that Flow*.

Finally, we want to thank the C&A Foundation, in the person of Douwe Jan Joustra, who has financially supported our work on the research, writing and design of this publication.

circular
business
models
and design
strategies

THAT FLOW

for fast-
moving
consumer
goods

CONTENTS

Preface

For a large part, our economy thrives on fast-moving consumer goods. They're mass produced and inexpensive and have a short lifespan. People buy them, or just get them without asking, every day. We're talking food and drinks, packaging, hygiene products, toiletries, plastic cups and cutlery, medical disposables, fast fashion, gifts and gadgets. After brief use, sometimes mere seconds, we consider them as waste, which in the most preferable scenario winds up being composted, recycled or incinerated with some energy recovery. Unfortunately, a lot of all the waste leaks into the environment as well and ends up in our streets, parks, rivers and oceans.

Surely, we can do without some of these products. Nevertheless, most of them do serve a purpose. We need food, packaging, clothes, disposables and gifts. They belong to our way of living and sometimes even who we are. This implies that we should put an effort into finding better ways to deal with wastage, either by preventing it from occurring, or by much more coherent material flow management.

There is already is a book, entitled *Products that Last*, by Conny Bakker, Marcel den Hollander, Ed van Hinte and Yvo Zijlstra. The first step one can take when working with products for brief use is considering a lasting alternative, for which the aforementioned book offers ample insight.

When that is not an option, we can use a new framework for fast-moving consumer goods. Where end-of-life products are a resource for new ones, or where they are rendered harmless. This is where *Products that Flow* comes in.

In the first two chapters this book provides an insight into the circular economy and the nature of short-lived products, what kinds there are and how they can be characterised in comparison with lasting products against the background of circular concepts for business and design.

The third chapter delves deeper into the various ways in which products that flow can lose their value, or their sense of potential. They vary from a very fast and immediately value drop, to very slow postponed rejection, like an unworn blouse hanging on a coat hanger for years.

Chapter three addresses business opportunities in and around products that flow. They're all about logistics, subdivided into organising, transporting and processing the flow of short-use products. In the next part the focus is on design for flow, which concerns themes that may also concern sustainability for lasting products. Chapter five offers a view on all the different families of materials. The last chapter focuses on us humans and what we already do, and on new developments that are likely to contribute to a more sustainable future. We are starting to discover common ground.

Siem Haffmans – Marjolein van Gelder – Ed van Hinte – Yvo Zijlstra

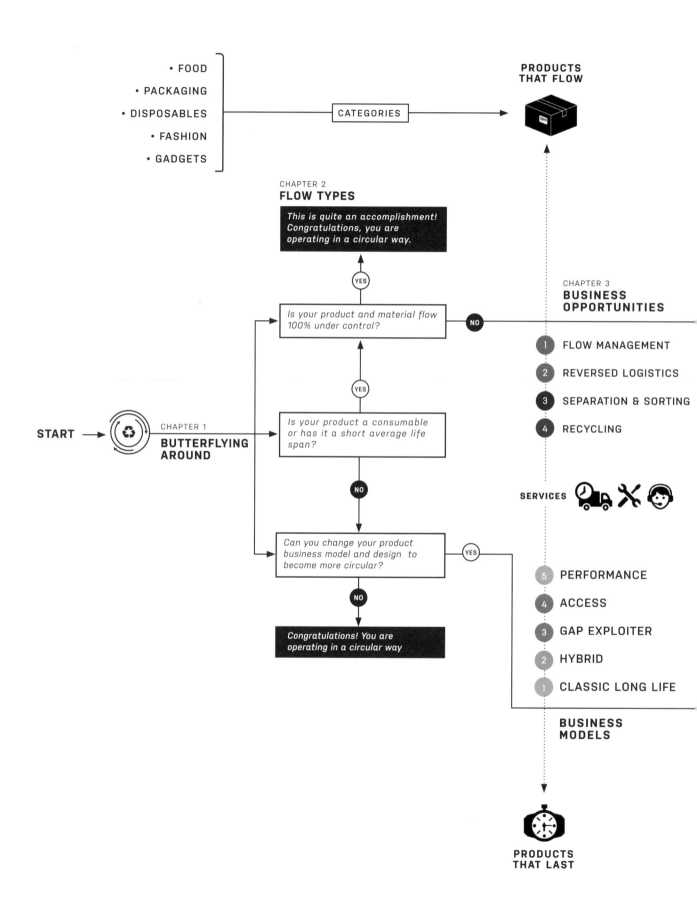

- FOOD
- PACKAGING
- DISPOSABLES
- FASHION
- GADGETS

CATEGORIES

**PRODUCTS
THAT FLOW**

CHAPTER 2
FLOW TYPES

*This is quite an accomplishment!
Congratulations, you are
operating in a circular way.*

YES

CHAPTER 3
**BUSINESS
OPPORTUNITIES**

*Is your product and material flow
100% under control?*

NO

1 FLOW MANAGEMENT

2 REVERSED LOGISTICS

YES

3 SEPARATION & SORTING

4 RECYCLING

START →

CHAPTER 1
**BUTTERFLYING
AROUND**

*Is your product a consumable
or has it a short average life
span?*

NO

SERVICES

*Can you change your product
business model and design to
become more circular?*

YES

5 PERFORMANCE

NO

4 ACCESS

3 GAP EXPLOITER

*Congratulations! You are
operating in a circular way*

2 HYBRID

1 CLASSIC LONG LIFE

**BUSINESS
MODELS**

**PRODUCTS
THAT LAST**

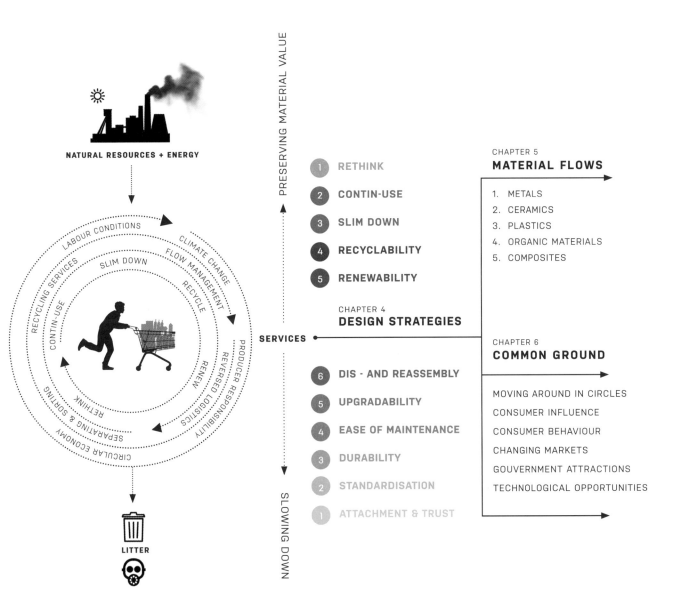

NATURAL RESOURCES + ENERGY

PRESERVING MATERIAL VALUE

SLOWING DOWN

LABOUR CONDITIONS

CLIMATE CHANGE

FLOW MANAGEMENT

RECYCLE

RENEW

REVERSED LOGISTICS

PRODUCER RESPONSIBILITY

RETHINK

SEPARATING & SORTING

CIRCULAR ECONOMY

CONTIN-USE

RECYCLING SERVICES

SLIM DOWN

LITTER

SERVICES

1 RETHINK

2 CONTIN-USE

3 SLIM DOWN

4 RECYCLABILITY

5 RENEWABILITY

CHAPTER 4
DESIGN STRATEGIES

6 DIS - AND REASSEMBLY

5 UPGRADABILITY

4 EASE OF MAINTENANCE

3 DURABILITY

2 STANDARDISATION

1 ATTACHMENT & TRUST

CHAPTER 5
MATERIAL FLOWS

1. METALS
2. CERAMICS
3. PLASTICS
4. ORGANIC MATERIALS
5. COMPOSITES

CHAPTER 6
COMMON GROUND

MOVING AROUND IN CIRCLES

CONSUMER INFLUENCE

CONSUMER BEHAVIOUR

CHANGING MARKETS

GOUVERNMENT ATTRACTIONS

TECHNOLOGICAL OPPORTUNITIES

The invisibles

So excellent are we as humans in production and consumption that in the recent past we have tended to overlook the large-scale consequences of our material consumption. These days we are becoming aware of the damage we cause, and we are gradually learning to be more careful.

VALUE DROPPERS

The foundation of improvement is understanding precisely what we are doing, in order to be able to develop and execute appropriate countermeasures. There is always a tendency towards imprecision, caused by the expectation that there will be some sort of simple fix behind a single idea with an appealing name.

However, things are more complicated. Products that 'flow' are being produced, stored, used, processed, and turned into an amount of waste that is often uncontrollable, but their use is relatively limited in time. For these 'products that flow', further exploration of ways to render deployment less harmful is urgently needed. A first important issue reveals itself already: we should start counting and measuring accurately what happens to the stuff we produce, to be able to assess the effectiveness of counteractions.

ECONOMIC BACKDROP

The whole of our behaviour and the way in which we humans organise our lives, is usually described in a set of household models that we call economics. Models are under constant development and some basic assumptions currently are under scrutiny. The first one of course is that a human is a rational being that invariably takes well-considered conscious decisions. We now know that this is not the case. There is no *economic man*. This is a hard nut to crack, for how does one deal with irrationality in rational ways?

The second assumption, that economic growth is essential for welfare, is challenged by an economic ideal created and promoted by English economist Kate Raworth. It is called 'Doughnut Economics' and, without going into all the details, it advocates balanced economics, whereby everybody lives

Products that flow are supposed to performe just once. They serve to prevent damage during transport, preserve quality, add convenience and seek attention to promote sales. Most end their existence as litter the moment their job is done.

on an acceptable level of social wellbeing and technological side-effects, such as pollution and climate change, are under control. Raworth proposes her system as an alternative to one based on economic growth, which simply cannot be sustained because consumption can only grow at the cost of resources and energy.

Kate Raworth introduced the Doughut Economics model as an alternative to the unsustainable growth model.

This points to an extra 'embedded' assumption, which is that economic growth is driven by material consumption. Since fundamental change is needed to stop the exhaustion caused by our ongoing craving for objects and stuff, it could help if growth were defined in terms of quality of life rather than sheer volume of the production consumption cycle.

Based on this third assumption, that we should think in cycles, a more concrete economic model has gained considerable popularity over the past years: Circular Economics. It describes an optimum economic reality, with natural cycles of matter as a metaphor: everything should go around and around, and be used, reused, refurbished, remanufactured, repaired and recycled. Hereby we distinguish between a renewable cycle, with organic materials that we can grow, and a technical cycle, with metals, ceramics and synthetic materials. Although the model is not perfect, it does provide principles and insights to gain control over both lasting and flowing products.

FLOW TO LAST

It is not as if some products exclusively 'last' and others strictly 'flow', like the book titles 'Products that Flow' and 'Products that Last' (2014), seem

Fast-selling fashion products with ever-shortening lifespans; products that used to last that start flowing.

to be suggesting. As a matter of fact, almost all products flow to some extent. They are being produced and transported, sold and used, and sooner or later discarded and demolished, or recycled. There are very few exceptions. Some end up in museums. Survivors may take ages to really disappear. They will certainly include some products that flow, such as throw-away lighters.

Yet, a striking difference between the two kinds of products justifies this book: whereas products that last are supposed to be able to do this, products that flow simply are not meant to last (maybe wrongfully)

and need a different approach because of this. They are supposed to perform just once, for a short period of time, maybe just for commercial seduction, although many flow products, such as fashion items and decorative objects, may look like they are made to stay. This does not imply an immediate end to their existence after they've done their job. They keep on requiring effort through transportation and processing (which defines flow), to either help them disappear altogether or to return them to a certain level of usefulness. Currently for a large part they are abandoned to their fate.

Once flowing products have achieved their aim, they quickly drop in value. This word 'value' is more often used than it is understood. Value is not necessarily expressed in money. In general, there are two opposing psychological 'directions'. It can either be a 'sense of oughtness', a feeling of how one should behave or how things should be, and, on the other hand, it can be a 'sense of potential'. This is an idea of what it can do later on and to what extent, either positively, or negatively.

Oughtness might, for instance, say to you: 'Put this in a bin'. Potential might call out: 'Hey, I am empty, but you could fill me with lemonade', or in the negative case: 'Don't touch me, your fingers will get all sticky'.

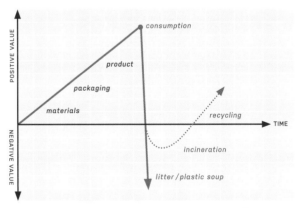

The typical steep value drop of products that flow

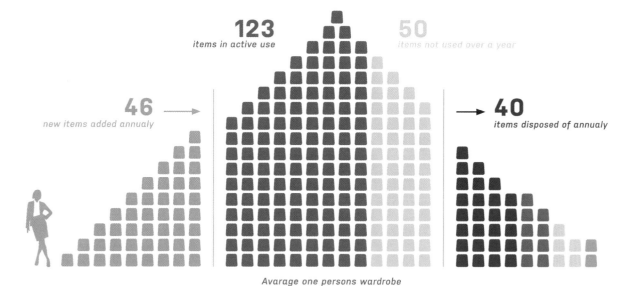

123 items in active use

50 items not used over a year

46 → new items added annualy

→ **40** items disposed of annualy

Avarage one persons wardrobe

On avarage the Dutch have 173 items in their wardrobe of which over 30% is not in active use. Each year the Dutch buy more and throw away more, yet the wardrobes keeps expanding. (Amsterdam University of Applied Sciences)

On the individual level, values continuously change. Aggregation of individually perceived values turns them into social values, which change much slower. If negotiations and transactions are involved, social value turns monetary.

Oughtness of cleaning streets cannot happen for free. But the cleaner, who may be a parent and a friend and a pool player and a prankster, represents value far beyond mere cleaning functionality.

Transactional value is in fact, currently, far more exceptional than the discipline of economics leads us to believe. There is much more to life and its arrangements than moving numbers between digital servers.

The mentioned sudden drop in value for some objects, for instance a cardboard box as soon as its content is freed, simply is a decrease in perceived potential. Oughtness may serve as a stimulus to do something about the low potential waste, but then it might not. If all values are lost we have reached a state of negligence, which we now know doesn't achieve anything, except an increase in littering.

The sudden value drop needn't always be final. For an eaten apple or a used shampoo bottle it usually is. For an 'unbelievably cheap' blouse, however, or a cute purple fake watering can, the purpose of which is providing brief shopping fun or momentary gift admiration, value drops more gradually, in a process

that can best be described as 'suspended rejection'. The first value loss may happen quickly, but after that it may take years of oblivious episodes in the wardrobe or the attic, alternated with moments of rekindling consideration of possibilities – perhaps I could use it to, hmm, or maybe Bobby would like it as a toy, or maybe I should just put it in the bin. There are various kinds of flow that need to be mapped for better understanding.

Analysing principles of flow is essential to be able to make appropriate adjustments. In an abstract sense one can say that, since there are methods, business models and design strategies to make products last, the most obvious way to improve products with a

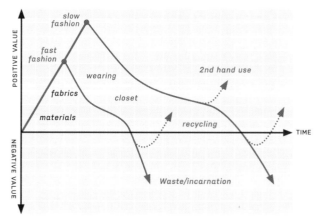

The value drop of garments

The supermarket is a combination of a logistic machine and a theatre where products that flow perform and seduce.

sudden value drop, is to start considering them as products that last, to circumvent their downfall as it were. For business models and design strategies this leads us back to the earlier book *Products that Last*.

There is another difference with the products described in this book: they are all complete products, each meant for consumers that use it. Products that flow are often temporary complements of other products, for instance when they serve as packaging.

DEALING WITH FLOW

When sustainability is required, to make all things last, however, is not necessarily the best option. Food is by definition impossible to keep in good shape forever. Sooner or later one digests it. Everlasting packaging can lead to grotesque amounts of storage and transportation needs and disposables are convenient and manageable if specific properties and conditions are required. Cheap clothing and house decoration objects may outlast the time that they are actually appreciated.

This implies that flow itself becomes the most important property of products to render harmless. It needs to be reduced from resources onward, controlled logistically and processed in ways that put a minimal burden on our living environment.

This is where the notions of open loop and closed loop and all possibilities in between become helpful, particularly if we define closed as 'fully controlled' and open as 'left for society to deal with'.

Since flow is so important, the 'product' can be a technological process that helps to close the loop.

Business models, or preferably business scenarios, because they evolve over time, are likely to have a character that differs from the ones discussed in *Products that Last*. It could concern a flow-enhancing business, or even a policy to enforce flow control.

Since reprocessing can be quite important, some design strategies will be similar to the classic ones, such as 'design for disassembly'. But they will have more refined consequences. Disassembly may be about complicated functional material compositions, the number of which is likely to increase anyway. Other design strategies may concern completeness of the flow itself: special features to accommodate packaging return. These are the areas where we expect true innovation to occur.

A SENSE OF OUGHTNESS

Noblesse oblige says that with wealth, power and prestige come responsibilities.

In ethical conversations, it can summarise a moral economy, in which privilege must be balanced by duty towards those who lack such privilege or who cannot perform such duty. Recently it primarily refers to public responsibilities of the rich, famous and powerful, notably to provide good examples of behaviour or to exceed minimal standards of decency. It also describes a person taking the blame for something in order to solve an issue or save someone else.

Public distrust of corporations has been rising for decades, and some suspicions came true. One of those is that multinational corporations have harshly outsourced production to poor countries. They force small local suppliers to produce at extremely low prices. They can only comply by employing workers, mostly women, against exploitative wages, in Ukraine as little as one tenth of living wages, working under subminimum safety conditions in sweatshops. Some prominent critics have called this modern slavery. Many are lobbying for implementing labour codes of conduct, for boycotting perceived exploitative corporations and for banning imports from countries with lax labour laws.

On the other hand. economists from across the political spectrum argue that multinationals have highly positive impacts on developing countries, where employees and employers have voluntarily agreed to a contract and when foreign affiliates are acting legally. Some companies have initiated enhancement programmes. The German clothing firm Kik, for instance, is doing its very best to improve labour conditions in Bangladesh, among other things by penalizing local factory surveyors who have been found to be lying.

Purchasing a cheap T-shirt from your local brand-name store may seem innocuous, but thousands of workers in the Far East, but also in southeast Europe are paying a far higher price than what is written on the tag. As global competition drives down wages, importers reap the profits of purchasing goods for a fraction of what they're worth.

In *Le Lys dans la Vallée*, written in 1835 and published in 1836 Honoré de Balzac recommends certain standards of behaviour to a young man, concluding: 'Everything I have just told you can be summarised by an old expression: noblesse oblige!' His advice included comments like 'others will respect you for detesting people who have done detestable things.'

1

shared concerns

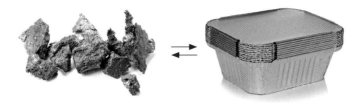

The most resonating sustainability buzzword today is Circular Economy. Like its predecessor Cradle to Cradle, it is considered to be the principle by which businesses can arrive at the solution to the all too well-known environmental mess we have got ourselves into.

BUTTERFLYING AROUND

The word 'Circular' is omnipresent and, apart from a few lonely sceptics, everybody wants to get involved in quickly circulating humanity's material processing towards a better future. Its representation of trade reality is not entirely adequate. Materials don't move around in circles and time goes on. Still, the Circular ideal, fuelled as it is by enthusiasm, can indeed serve as a kind of supercharger to help us make adjustments for a better world and to improve the model itself on the go.

CIRCULAR PRACTICE Usually the Circular Economy is depicted as the Butterfly diagram, for that is the shape in which the Ellen McArthur foundation promotes it. The 'left wing' of the butterfly is the circle of renewable or organic materials. Food, for people and animals, is an important component and many others, mostly trees and plants, contribute to all existing kinds of products one way or another: clothing, furniture, packaging, books, building and construction, ships and cars. Even airplanes have

wooden parts to make luxury class passengers feel at home. There is no trade that doesn't depend on renewables to some extent. They have the advantage that they just grow and that it takes no effort to break them down. Recycling just happens. Insects and micro-organisms do it for us. It may take energy, however, to separate them from other waste and to transport them to appropriate locations for digestion.

On the right side you will find the circular 'wing' of technical materials, which unjustly may suggest that organic materials are not subject to technical handling, but that is not the point.

The property that counts, is that these materials are originally mined, not harvested, and consequentially man-made, or man-composed, with the help of machinery and usually considerable amounts of heat.

Therefore, it also takes effort and energy to decompose them. There are three main types of

Photographer Gregg Segal pictured people of all ages and backgrounds, surrounded by a week's worth of their rubbish, including recyclables, for his series 'seven days of garbage'.

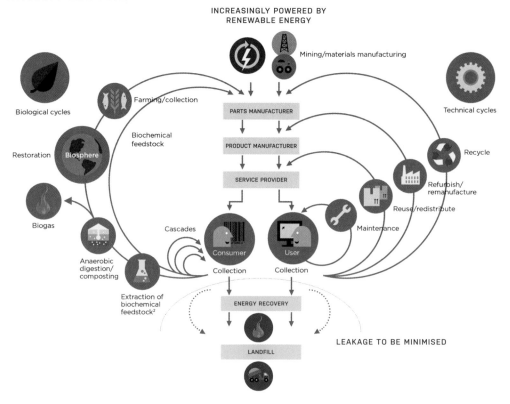

The Buterfly model (source Ellen MacArthur Foundation).

materials: ceramics, metals and plastics, which all exist in thousands of combinations, variations and physical states. It takes various degrees of effort to produce them and to keep them doing their job as long as necessary.

Look at the products that surround you and you will soon notice that the biological and the technical wing are not entirely separate universes. Even something as straightforward as a cardboard packaging box may have been combined with plastic parts and sticky tape, as well as metal staples. Products quite often consist of both kinds of material. This implies that the circles represented by wings are often intertwined and have to be separated when recycling is at hand. This is the simplest demonstration of the fact that the ability to recycle more than anything depends on logistics and flow control. Essentially all materials can be recycled, albeit not equally well, but they have to be identified and kept apart to be able to purposefully reprocess them.

LOOPS The word 'circular' indicates that products move around, without any input or output, like a perpetual mobile, except that it doesn't produce energy,

but consumes it. Materials and goods are harvested, mined, transported, and processed. Take a closer look at the diagram's top and you will see tiny arrows on the renewable left picturing an exchange of matter and on the technical right a supplement: the circle needs fresh nourishment. Replenishing depends on the kind of material involved. The better they can be recycled the less resources need to be used for fresh supplies.

At the bottom of the diagram there are larger arrows that illustrate removal of waste. Superfluous matter that may be useless or toxic needs to be dealt with. All this adds up to the observation that the circle is the ideal flow, but that it requires effort to make it work. We can approach it by making it less dependent on resource input, thereby diminishing the output too, slimming down the flow as it were. That is the only way to keep the output of waste to a minimum and to reduce the amount of energy required.

As it is, the circular economy can diminish use of resources, and postpone waste production. However, by itself it contains no incentive for flow reduction. At the end of the day, recycling needs waste, and since economic growth is, by default, connected to a

Decomposing or depositing: in both cases a system and a lot of energy for collection, transportation and distribution is needed.

continuous increase in consumption of materials and energy, recycling businesses thrive on growth of the amount of waste. It is appealing to look for ways to expand the effects of the circular economy by making it less dependent on fresh materials and energy.

Reaching the perfect flow of materials in the sense that no waste is produced, is often addressed as 'closing the loop'. This expression is a recycling cliché that has emerged from the current approach to sustainability, which is predominantly oriented towards engineering.

A wiser definition of the degree to which the loop is closed, is to what extent a particular organisation can be held responsible for waste.

Many breweries own their crates and bottles and they take them back for refilling through a deposit system. Currently, however, the handling of most waste is determined by the general public and government. Industry pays taxes, but individual companies are hardly responsible. As a consequence, a lot of waste 'is set free into the wild'. The loop is open. The keyword here is control. Closing the loop implies that for every contribution to 'the body of waste' it should become well established who's in charge.

MINIMUM ENERGY CONCEPTS Energy requirements are part of the deal and unavoidable. It is too easy to ignore them by assuming that sustainable power is bound to appear soon 'in an economy near you'. It is certainly true that a transition from fossil fuels and nuclear energy to wind and solar power is building up, with wind and solar electricity very slowly

replacing fossil fuels. Let us face it: the contribution of renewable energy to total consumption is small but encouraging. It is by no means certain that the current level of energy consumption can exclude fossil fuels within the near future. Some argue that the sun provides thousands of times the energy that the world population needs. They overlook the fact that sustainable power for all does not so much depend on availability, but rather on conversion (from radiation or wind or water movement into electricity), storage and transfer. Sure, the Gobi Desert is sunny, but providing large container ships with electricity is something else. These things don't happen overnight.

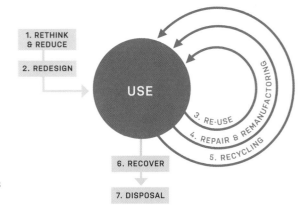

Think different: from size to value.

Shifting emphasis from replacing the generation of electricity by fossil fuels to reducing the need for energy consumption altogether, will speed up energy transition, or concretely: swapping a diesel crane with an electric one reduces CO_2 output, but hoisting lighter loads will take reduction at least one step further.

Another reason for attempting to use smaller amounts of energy is the uncertainty of change. The old idea that there would always be enough of everything should not be stretched to include sustainable energy production. A transition is on its way, but it is not entirely clear where it will take us. It is not just a matter of engineering clever systems. It depends on people. It depends on politics.

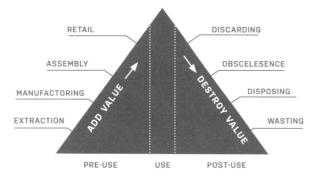

Product value is subject to continues change

GUARDING VALUE

The Butterfly Diagram represents the material flow as a circular sequence of essentially technological product interventions: maintenance, mending, milling, etc. What happens from the user point of view is more complicated: product value, as explained in the intro-duction, is subject to continuous change. For products that last, all effort should be directed at keeping up the value, or sense of potential, as long as required to reach the optimum lifespan ('Products that Last', chapter 'Ups and Downs'). This is the amount of time determined by the relationship between the gradual value change of the type of product and individual product value. A simple example: the optimum lifespan of a new car with a Diesel engine may be rather short and decreasing, because this type of engine will probably disappear. So, there is no point in cultivating its value, unless the car can easily be electrified.

This implies that ('Products that Last' chapter 'Round and Round it goes') a representation of the handling of long-lasting products, a mountain with a high value peak with value cultivation efforts around it, would be more appropriate than just a flat a circle, since the mountain would illustrate that effort is necessary to keep product value, or a sense of product potential, as high as possible.

Interestingly, this implies that recycling is not a particularly ideal treatment. It takes energy and it results in destruction of built-up value: tumbling down the mountain. To be more precise: in value terms, recycling is a fragmentation of a focused sense of potential, for instance a PET container for beverages, into a material with which some kind of new focus of potential needs to be created from scratch. It could become anything, from a bottle to filling for road building material, to a cardigan.

Value drop caused by sudden obselesence; the floppy disk

On the other hand, throughout their entire lifecycle, products and materials need to be optimally withheld from being harmful to the environment by keeping energy consumption, additional burdening of resources, pollution and emissions to a minimum level.

In value terms, a sense of negative potential, and an inclination to be harmful and to increase environmental costs, needs to be recognised and excluded as best we can.

The circular model serves as a qualitative benchmark. It helps establish where you are, when you decide to start a repair service for robotic dolls, or when you are considering to use recycled plastic packaging to produce furniture. An important overall observation is that flows are preferably clearly defined, to be able to control of the values of the products concerned. The less is known about purpose and what to do next, the lower the value, so much so, that it can be negative: you have to pay dearly. A case that is designed to be used for years on end to protect a costly cello, has easily sustainable value. At the bottom of the value hierarchy, we find garbage that didn't properly wind up in a bin is unlikely to ever return to dignified usefulness.

Go with the flow: Cultural ingeniuity leads to improvised musical instruments made from valueless scrap materials that enable them to produce extremely rich music, from the Junk Funk of Soweto to the Recycled Orchestra in Paraguay.

THE FLOW PERSPECTIVE

Looking more precisely at the circular model, we can make an observational distinction between product use and maintenance and all that comes with it in the centre, and processing, handling, storage and transportation, the actual flow, around it.

If we take, for instance, a washing machine as an example, then we can analyse the environmental effects of its use – including energy consumption- and maintenance and repair. In addition, we can have a separate look at its production and transportation, and the scrapping and recycling of materials that consequentially can be reworked to become parts for washing machines or whatever other objects: the flow part.

For a typical flow product – in this case it could be the laundry detergent and its packaging – sustained use is not likely. From the user point of view the soap is what matters and the plastic container is a mere backstage prop. It is necessary to hold the soap and it provides a bit of branding information, but it is not an object the receiver ever consciously asked for. Opening and closing the bottle is the only interaction. Consequentially, the user could either discard it for processing as waste, or continue using it, maybe to water plants. Currently the former way of dealing with it is common, although the container may be temporarily saved to store things, or as a toy for kids or dogs. It depends on circumstances, design and financial considerations, Whether or not reassigning it to its original packaging that deployed it depends on circumstances, design and financial considerations.

Packaging is just one category of products that hardly boast an episode of conscious use. Others have different characteristics. Some are indeed meant for consumers, to be literally consumed, like cauliflower or chocolate, or put in a vase, like tulips, or applied, like shampoo. Others, such as cheap tops and blouses, are meant for consumers, who buy them, but scarcely wear them. Some items can be said to be an excuse for undergoing the always enthralling shopping experience.

No matter what, products that flow participate in games that are considerably different from the ones that lasting products take part in. Whereas products that last are always meant for a client who is also the user, the client for packaging is usually the company that sells the product to be packed. There are also companies that do part of the processing for a producer, such as printers. Or they may exploit a truck fleet for transportation. Products that are meant to be consumed directly, such as fashion products, gadgets, knickknacks, or giftware, are mainly put on the market to be sold, or to communicate symbolic value to those who receive the present. Actual convenience comes second. Food is part of yet another flow type altogether. This implies that the world of flow, which is gargantuan, is rather blurred and that it defines a different, less familiar view on businesses than the world of durables.

The overall principle is that flow is the main concern for the products involved. Flow effects should be minimised.

We have a turnaround of the notion of value on our hands: the potential of flow to minimise detrimental side effects increases flow value. Optimisation of various types of flow, is the challenge faced by businesses and designers.

DISPOSABLE IDENTITIES

BIODIGRADABLE SUNGLASSES
The eyewear series Collection 1, developed by the studio Crafting Plastics in Berlin, is an example of managing the lifecycle of fashion products, particularly after they go out of style or are discarded. The material used for the frame is PLA (Polylactid acid), a polyester plastic that can be moulded when heated. It is produced from different plant-based substances, such as corn starch, tapioca or sugar cane. The plastic technology has not developed far enough to be able to mass produce lenses yet.

PAPER FASHION
Photographers Alexandra Zaharova & Ilya Plotnikov in Moscow experiment with paper fashion for photoshoots. Not a very practical material to defy the elements yet but considering that cardboard once replaced wooden transport boxes, paper may offer unpredictable applications in the near future.

SPRAYED FASHION
Thanks to a liquid cotton fibre mixture, we could soon be spraying ourselves into everything from T-shirts, dresses and trousers to swimwear and hats. This technology, called Fabrican, is based on 15 years of research by fashion designer Manel Torres and particle engineer Paul Luckham. When you get bored with your creation it can be washed off and dissolved and the material can be used again to make something new or to repair or redesign worn clothing.

UPCYCLING
Electrical wiring transformed into jewellery and Tyvek shopping bags transformed into a raincoat.

NO NEED TO BUY
One in seven women photograph themselves trying on new outfits, post it to social media, and wait for an average of two likes from friends before they decide to buy. Many even take the opportunity to get images of themselves in various fancy outfits with no intention of purchasing. Sounds like a new business opportunity.

BIODIGRADABLE SHOES
Adidas may have come up with a biodegradable shoe. Called the Adidas Futurecraft Biofabric, the company's new shoes can break down in under 36 hours after you add a special enzyme. The shoes are made with an ultra-strong, lightweight material called Biosteel, produced by biotech company AMSilk. This knitted fabric, which forms the shoe's upper, is created using the same proteins that spiders use to make their web.

stories for grabs

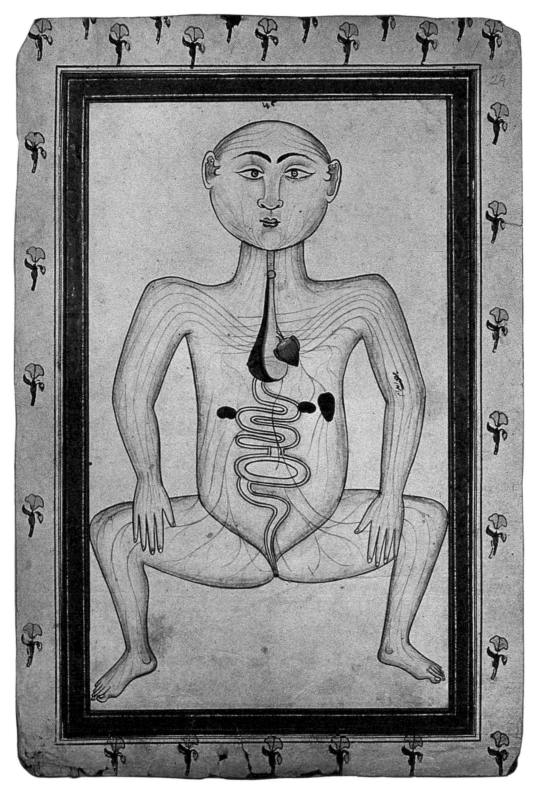

Persian anatomical figure (1930) showing our digestive system and flow types with energy going in and waste going out while efficiently closing the loops, avoiding deregulation of the system through abundance or poisoning.

The flow of energy: batteries are to be recharged and ignored to support our way of living until they die.

Products flow in different ways. There are several categories of course. Both between and within those, the speeds and durations of value change are different. For instance, in the category of disposables, the 'lifestyle' of a disposable surgical knife will totally deviate from the experience a disposable battery goes through from the factory to the recycling plant, which in turn has no similarity with a plastic fork. Their respective value behaviours are different.

FLOW TYPES

The chosen subdivision of products that flow is rather straightforward. What they share is, apart from their inclination to quickly lose attention and value, the fact that they have very little in common. Every trade has something to do with packaging, but only packaging people have seen the strings. Essentially each category represents different trades, different universes even. A producer of gifts doesn't know what is going on in the world of food. Putting the trades together in one book may provide valuable cross-linking opportunities.

Food flows like no other product. We farm it, harvest, process, transport and store it and finally consumers, including us, buy it and eat it. This so-called food chain is evolving. The average American consumed 3641 calories a day in 2011, which is about 26 percent more than in 1961. The increase

of calorie consumption during the same period in the UK was fifteen percent and in Germany eighteen percent. As a consequence, worldwide obesity has taken over from famine as the most important food problem.

Eating too much and an unbalanced diet is wastage. Even with our metabolism a staggering 30 percent of the food produced worldwide remains unconsumed. Americans even throw away almost as much as they eat.

There are several causes for this. One is looks. Selection during and after harvest renders fruits and vegetables unsalable due to lack of attractiveness. A second reason is the sell by date. Products are no longer sold after that and usually are useually discarded. Yet, nothing is wrong with them, except perhaps for a slight decrease in quality. As a matter

A perpetual stew is a pot into which whatever one can find is placed and cooked. The pot is never emptied all the way and ingredients and liquid are replenished as necessary.

Proper protective food packaging for transport is crucial. Wrong packaging can cause loss of total food production due to damage and rotting.

of fact, an increasing number of restaurants now specialise in meals prepared from 'overdue' ingredients. The use by date differs from the sell by date. After that the product, such as fresh milk or fruit, gradually becomes inedible. Consumers are not sufficiently aware of this difference in dates. Moreover, they purchase and serve more than their guts can digest. In food culture enough is always more than enough and never by just enough. Socially, consumption is driven by hospitality, which often turns out to be misplaced generosity. The *All you can eat restaurant* is not really a good idea.

Apart from losses in other sections of the food flow, consumers in the Netherlands on average throw away 41 kilograms of food each year. Since 2009, the Netherlands have pursued a policy against wasting food. Nevertheless, the amount of food waste remains more or less the same. Accurate data are lacking, but the target of 20 percent reduction by the end of 2015

was not achieved, despite the fact that many Dutch people say they reject wastage. In fact, consumer expenses for food in the Netherlands are increasing.

Agriculture and food have rather different characteristics than the other categories, since flow is brutally interrupted by preparation, eating and digestion. For that reason, they will not be analysed extensively in this book. Even so, there is likely to be a mutual associative effect.

Packaging is the most present and probably also the most neglected kind of products that flow. In most cases the packaging, which is subordinate to whatever is inside, is used just once, with a one-way ticket to the bin after that. When using a package only once before we throw it away, its value gets lost in a short period of time. Most plastic packaging is used only once, thereby immediately losing 95 percent of its value. Annual costs are estimated at USD 80-120 billion.

Tissue paper slows the rate of rotting, retains O2, and also prevents neighboring oranges from spoiling one another. The cost to the industry can run into millions of dollars and eco-friendly consumers are against the waste paper that is produced.

A well filled food store inside Ben Thanh Market in Ho Chi Minh City, Vietnam

By no means does this imply that packaging is pointless. Packaging first and foremost serves to protect its content against mechanical damage (during transportation and storage), the influence of moisture, heat, (UV) light, disturbance by bugs or other organisms and it can maintain the quality of the atmosphere inside, for fruit and vegetables. In Western Europe, 2 percent of food waste relates to unsuitable packaging, while in African countries these numbers can be as high as 30 to 50 percent.

Moreover, packaging can serve to hamper shoplifting, particularly when it concerns small but valuable electronics: just put them in an unbreakable large plastic envelope and you're home free.

Next of course packaging may carry information, such as 'this side up', or 'fragile' and content specifications with a serial number. Information can also function as a manual and of course contribute to advertising and branding. This is part of the whole design. The experience of opening a package can reassure the client who bought the object of desire in the wonderful box. The purpose of packaging should always be weighed against its environmental burden. Packaging has to do what packaging has to do.

The packaging market is prosperous. The fastest market increase occurs in Asia-Pacific countries and other regions of the world where economies are growing rapidly through industrialisation, urbanisation, growing population, increasing international trade and of course ecommerce. Legislation, concerning safety, labelling, transportation and storage, helps as well.

A specific phenomenon is on the increase in emerging markets: small sachets. Though these were originally meant for the poor, who could only afford small portions that they bought in street shops, currently Asian supermarkets also sell them in bulk. For example, more than 70 percent of the shampoo sold in Asia nowadays comes in sachets. Unfortunately, they are difficult to recycle, since they consist of laminated foil. Moreover, they are too small to collect by hand: there is no business in them.

At the moment, worldwide only fourteen percent of packaging is collected for recycling. A subsequent four percent of this amount is lost in the process, eight percent is used in lower quality applications and only two percent of this fourteen percent actually makes it to closed loop recycling.

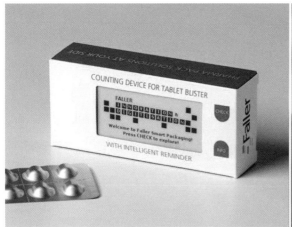

Smart packaging contains integrated technologies, such as NFC chips, Bluetooth, LEDs or screens for food and drugs, reducing potential waste throug better stock management.

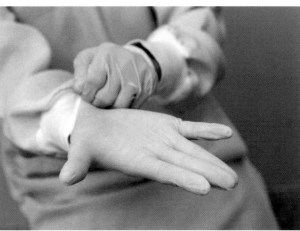

Biodegradable disposable gloves for medical use and food industries. In 2017 the global rubber gloves market was valued at 2.35 billion U.S. dollars.

Systemic changes are required to change the way we pack and deliver our products. This will take time. However, there are also readymade activities that companies may be able to start implementing right now. They involve measures such as rethinking the system in the direction of sustained use of packaging, or a switch towards biodegradable materials.

Disposables' sales are on their way up too. There are two main areas of disposable trade. Medical disposables are used in surgery, drug and fluid administration, wound management, diagnostic testing, patient examination, incontinence management, dialysis treatment, sterilisation and blood donation etc. And finally, infectious waste disposal. The turnover in medical disposables is getting larger, due in part to an ageing population, improved health facilities, but mainly because handling, packaging and management of cleaning and sterilising used equipment, instruments and fabrics has become unaffordable. Processing rules for sterilisation are very strict.

The average amount of disposables used in the USA is 14.9 kilograms of waste per patient bed per day, which adds up to 2.6 million kilograms of waste each year.

There could be a solution in continued use for parts of the equipment, when, for instance, discarded instruments and sheets are cleaned, but not necessarily sterilizsed anymore, for normal unsterile use. Currently this happens informally.

Demand for foodservice disposables in the USA is growing too. It is a multibillion dollar trade expected to grow at a rate of several percent per year. Limited service restaurants, which account for 80 percent of the market, push most of the growth.

Retailers also continue to expand their range of prepared foods in order to better compete with restaurants and takeaway brands. In addition, restaurants add new beverages to the menu, such as coffees, smoothies and teas. These changes of course boost the introduction of new food disposables and packaging. The challenge to find harmless alternatives is formidable, and captivating because of this.

Fashion drastically changed in the 1990s. It used to be based on certain designers proclaiming looks and silhouettes for the coming year and the market purchasing accordingly, well sort of. Now it has developed into the most destructive trade after the fuel industry. Style change has almost come to a standstill, but items with 'different' details, prints, folds and seams, evolve from computer drawings to the shop shelf in a matter of weeks. Items are produced very cheaply under harsh circumstances in the Far East, but also in countries like Romania, Albania and the Ukraine.

The items are so cheap that customers often don't even bother to try them on in the fitting room. Clothes are only worn a few times, if at all: of 173 clothing items in the personal wardrobe in Denmark and the UK, about 50 have not been worn over the past year. At times items are disposed of to create wardrobe space for new identity illusions. The young

Biodegradable foodservice products are becoming increasingly popular with business owners who need to meet the wishes of their customers in looking for new ways to protect the environment. The best way to reduce waste is to not produce it in the first place. This is often called precycling or source reduction. Removing the trays so students do not take too much food has reduced campus dining facilities food waste by around 35%, saving millions of gallons of water and chemicals and cutting labour costs for dishwashing.

Samples from thousands of cheap apperal items filling a popular European online fashin sales website.

generation in particular aches to own a large amount of clothes to don and play around with, making selfies and putting them on social media.

The size of the fashion market is estimated in trillions rather than billions of dollars and expected to grow rapidly. Apparel consumption is projected to reach over 100 million tons in 2030.

Obviously, production of this volume of new clothes has a significant environmental impact. The industry mostly relies on non-renewable resources: close to a million tonnes per year. These resources serve to produce synthetic fibres, fertilisers to grow cotton, and chemicals to produce, dye, and finish fibres and fabrics. Chemicals for treatments like dyeing account for twenty percent of the global industrial water pollution.

The total greenhouse gas emissions from textile production exceed those of all international flights and maritime shipping combined. Moreover, annual water consumption for textile production is around 93 billion cubic metres: a challenge for the local ecosystems. Chemicals are widely used in the clothing production industry, for example for the dyeing and treatment of the fibres.

Waste from the fashion industry (production to end-of-use) is predicted to increase by about 60 percent by 2030 totalling annual waste of around 17.5 kilograms per capita across the planet. Currently, most of clothing waste ends up in landfills or is incinerated. Globally, only twenty percent of clothing is collected for continued use or recycling, but actual recycling hardly happens, simply because it is difficult to sustain material value, particularly over several reprocessing cycles.

Luckily, there are opportunities to develop a less harmful fashion system. According to the Ellen MacArthur Foundation, the textile industry should pursue four objectives: phase-out microfibre release,

2.6%
The percentage of global water used for growing cotton.

17-20%
The estimated percentage of industrial water pollution that comes from textile dying and treatment.

60,000,000
The estimated number of people who work in the fashion industry worldwide.

700 GALLONS
The amount of water it takes to produce a single cotton T-shirt.

8,000
The estimated number of synthetic chemicals that are used worldwide to turn raw materials into textiles.

68 POUNDS
The amount of clothing that the average American discards each year, 85% of which ends up in landfills or incinerators.

99%
The estimated percentage of used clothing that is recyclable.

Some indications of the use of resources and global environmental impact of the fashion industry.

increase cost utilisation, make radically more efficient use of resources and a move to renewable inputs. In addition, fashion doesn't have to be made from textiles. Identity expression doesn't necessarily depend on selling products. Rules for product lifespan in relation to environmental load are required.

Some measures are pretty straightforward. Reduced water consumption could save up to 32 billion euros worldwide by 2030. There is a possible 160 billion euros per-year to be gained for the world economy through more efficient and diligent use of scarce resources in the apparel and footwear industry.

In general, a lot is to be gained from making progress on a range of issues all along the value chain and treating textile workers fairly and providing them with safe working conditions.

Gadgets and giftware are a rather unusual category to describe. Not much general knowledge, about volumes and effects and production circumstances, is available. There are two reasons for this. Firstly, gad-

gets and giftware cover a wide range of different products, from tiny travelling irons to coffee table books and from fluffy toys to flower arrangements. Secondly, they succeed in escaping the attention of designers and producers, which may be the very reason they need to be discussed: nobody is really bothered. Therefore, their value change is similar to what happens to fashion items, the main difference being that most of them don't live in wardrobes, but rather in full sight, on window sills, on shelves, or in gardens.

FLOW EPISODES As we mentioned in the beginning, products that flow are characterised by a moderate idea of potential that diminishes fairly quickly in the beginning, although speeds may vary. This value drop in some instances is followed by doubt if the product should be kept and used, or discarded. At the end it is usually discarded and after that it winds up in a biological or a technical renewal process, or it is burnt, or it ends up in the nowhere land of unorganised waste. What happens precisely depends on the kind of product and its context.

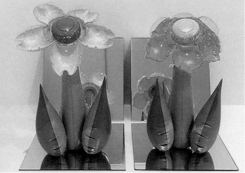

In the popular imagination, balloons often represent freedom and joy. Nevertheless, there is a profound disillusionment that accompanies the release of a balloon, namely, what goes up must come down, often with a loud pop. The same goes for inflatable flowers celebrating nature, unless they are from the limited edition series 'Inflatable Flowers' from 1979 by American artist Jeff Koons. His handmades have a deceptively similar appearance to mass produced goods.

Value change; the fleamarket as the stock exchange for both lifetime extension and suspended rejection.

An overview of these so-called *value scenarios* for a range of different products can be helpful to decide on business adaptations and design strategies. This is a matter of careful analysis and cross referencing.

To a certain extent it is possible to estimate and map value scenarios for flowing products. Up until now there have been no precise registrations of value change, but since this way of looking at product flow may point out directions to diminish waste, by comparing flows and by understanding what actually happens, it may become subject to scientific research. For now, the presented overview is speculative, but by no means implausible. Recognition is the aim.

For every kind of 'value dropper', we can – the outlines have already been crudely sketched – distinguish between three stages of value change. The first is what could be named the '*drop stage*', in which value starts fairly high and is entirely lost very quickly, often beyond rock bottom, where it costs money to remove the materials involved. This can happen in a matter of seconds. A piece of foil torn off from a candy bar loses all value instantaneously, unless it is used

for holding the sweet for some brief extra moments. Value loss can also take more time, say an hour up to several days for a pointless birthday gift 'that you always wanted' or something you bought because you thought it was a good idea at the time.

The next stage that was mentioned in the opening chapter is '*suspended rejection*'. This can be virtually absent, as in the previous case, but it can also take several years: you bought this seemingly appealing pair of trainers that you wore once or twice and then left in your wardrobe. They cause you to doubt whether you should keep the damned shoes whenever they accidentally catch your eye.

The final '*loop assessment*' stage concerns to what extent the product has an organised future, or to what extent 'the loop is closed'. In the current stage of sustainability development, it is the second half of what we're usually dealing with, which is gaining control of material flows and rendering them harmless in as far as they are uncontrollable. This used to be named 'end of pipe', wipe away the soot after production. The first half concerns prevention of value loss: start with a PET bottle, or a newspaper and a well-designed control scenario, and closing the loop is a piece of cake. Organisation is the holy grail.

FIFTEEN EXAMPLES We distinguish between the five types of products described earlier. For every category we provide three examples. They're bound to be familiar. The list is not intended to be complete. It is supposed to demonstrate a meaningful range of referential value scenarios.

Food and consumables are very fast value droppers

1.  RICE needs to be cooked and prepared to bring it to edibility level. Fresh rice may wind up in suspended rejection, if, for instance, it differs from your customary preferred brand. Usually the prepared rice is eaten for the most part. Leftovers have very little value. They become an ingredient for animal nutrition, they are composted, burnt, or they end up in a landfill. The part that gets eaten takes the sewer trajectory.

    FLOW TIME
   ORGANIC WASTE
   PLASTIC WASTE

2.  A CANDY BAR is usually eaten completely and takes the intestine bypass to the sewers. The only leftover is the wrapper: see packaging.

    FLOW TIME
   ORGANIC WASTE
   PLASTIC WASTE

3.  A BOTTLE OF SHAMPOO will lose a bit of value when it is opened for the first time. After that, value will degrade in a matter of weeks, according to the amount of shampoo used. You could say that the more soap is used up, the more suspended rejection of the bottle takes over. After this it may take days to decide if the bottle can be considered sufficiently empty to make it to the garbage bin.

   FLOW TIME
   ORGANIC WASTE
   PLASTIC WASTE

Packaging belongs to a product, until it is opened. Then it turns into waste in seconds. There are exceptions, but not many.

4.  WRAPPERS have been mentioned several times. They are the obvious example. No one cares about the foil around their ice cream, except perhaps for the odd ice cream wrapper collector. It is an immediate reject and typifies the kind of waste that litters landscapes. Its destiny: ovens, if some control is exerted. Otherwise anywhere.

    *FLOW TIME*
   ORGANIC WASTE
   PLASTIC WASTE

5.  HEADSET PACKAGING, like many other electronic commodity protectors, usually consists of a combination of parts: a rectangular cardboard box, or simply a card, plus a preformed plastic shape that neatly fits around the object. The combination more often than not is inconvenient to tear apart, so the whole thing is just waste, although in some cases one may keep them for a while because it seems to promise further use, until the decision to get rid of it turns unavoidable.

   FLOW TIME
   ORGANIC WASTE
    MIXED WASTE

6.  WATER BOTTLES are notorious. The approach differs per country. In the tropics, drinking from the tap may be hazardous and poverty renders the bottles not entirely without value for people who collect and sell them. In rich countries value may be kept up artificially with a deposit system. This is a reasonably accepted way to keep bottles within a flow of products that can be recycled.

   FLOW TIME
   ORGANIC WASTE
    PLASTIC WASTE

Disposables as a category contains many different value scenarios, because the number of different disposables is huge. They range from colourful plastic nothings for children's parties, to sophisticated surgical tools.

7. **CARDBOARD AND PLASTIC COFFEE MUGS** are used by the billions. They have very little value after use. There are just too many of them. Some are collected as a separate flow, which to a certain extent are recycled in the regular paper flow and the less well-defined plastic flow. The materials do not regain their original quality and proceed with a lower value. Many of them have PS (polystyrene) lids. PS, like all thermoplastics, can be recycled, but logistically the process is relatively expensive.

▏▏▏▏ *FLOW TIME*
▬▬ ORGANIC WASTE
▬▬▬▬ MIXED WASTE

8.  The value of **DISPOSABLE SURGICAL NAIL NIPPERS**, after ritually opening the package, is defined by sterility for their one-time use. When sterility is not required they are still useful. It is not uncommon for hospital personnel to take this kind of instruments home, such as scissors, or tweezers. Occasionally they are collected to be used in countries that cannot afford high-end sterilising procedures. They can last a lifetime and often do, without any episode of suspended rejection. There are of course many different kinds of instruments. They are not likely to all live by the same value scenario, but it can make sense to start looking at them from a sustainability angle.

▏▏▏▏▏▏▏▏▏▏ *FLOW TIME*
0
▬▬▬▬ METAL WASTE

9.  **SINGLE USE ELECTRICAL BATTERIES** lose value at the same rate as power. They do so until they reach the suspended rejection stage that will lead to replacement. Most of them will not have been totally depleted. The leftover energy is not salvaged. In this stage of the development of recycling legislation and processing systems, in the west about half of all batteries are recycled. This share is likely to increase. Moreover, the number of rechargeable batteries is growing, perhaps pushing batteries to the area of products that last.

▏▏▏▏▏▏▏▏▏▏▏ *FLOW TIME*
0
▬▬▬▬ MIXED WASTE

Fashion products, more than anything, serve to celebrate the idea of the self. In some more than in others and to a degree that decreases with age, self-image is continuously subject to experimentation. The purchase experience is a kind of self-search test.

10.  The value of **GARMENTS** can drop fast even the first time it is worn. Where blue jeans have been around for ages and may be worn for a long time (wear is part of their identity), a blouse that is combined with it may turn out to be a total identity failure. A parent's remark – 'you do look lovely in that shirt!' – can do the trick of tainting it forever. Afterwards it will find a place in between many clothes, some of which are cherished, others which are worn regularly, and some of which are cursed but kept for later consideration. Once rejected, clothes are difficult to deal with and tend to simmer on the down low. A small portion can continue as second-hand clothing or for sharing purposes. Un-weaving is expensive. Recycling hurts quality. It could be an idea to develop a section of clothing industry that is less hung up on textiles, such as nonwovens or new kinds of paper.

▏▏▏▏▏▏▏▏▏ *FLOW TIME*
▬▬ ORGANIC WASTE
▪ MIXED WASTE

11. **TRAINERS** are part of a massive complicated market. Some are purchased as an investment and never worn in order to keep them in their most valuable state. Many, let us say the mid-range ones, get an opportunity to simply do their job as products that last. A number of them are also cheap but seductive. They take the cheap blouse trajectory. There are attempts in favour of increased recycling potential. In the technical sense trainers are complicated objects, which doesn't help.

FLOW TIME
ORGANIC WASTE
PLASTIC WASTE

12. **CHEAP BRACELETS** are never really discussed in relation to recycling on the large scale that jewellery is produced and sold. They provide the same kind of purchasing experience as other fast fashion items, but they may be less difficult to mill and separate, being mere metal, glass and plastic.

FLOW TIME
0
MIXED WASTE

Gadgets and giftware usually remain undiscussed, because these products tend to stay under the radar of those who are interested in sustainable ways of living. But decoration and playfulness are part of who we are and they shouldn't be neglected.

13. **MINIPRESSO** has everything you wish for in a gadget. It is a little hand pump to press hot water through coffee. You can take it on your travels. It works fine the first time. You can demonstrate it to a friend and it usually appears not as practical as it seems. Rather quickly it will end up in a kitchen drawer to await further use, which will not happen. If all of its parts consist of the same plastic, recycling after its stay in the drawer is feasible.

FLOW TIME
0
MIXED WASTE

14.  **SOUVENIRS**, or objects that remind one of memories of old objects, like fake retro typewriters you can buy in knickknack shops or on the Internet, are just two kinds of many that serve as decoration, on mantelpieces, windowsills or in glass cabinets. People buy them because they find them charming and serve to help compose a personal idea of beauty and cosiness. Since they are literally useless they have the potential to last very long. On death row. The decision to discard such an object can happen anytime, for instance when a replacement bauble turns up. The discarded nostalgia carrier may be very expensive to recycle. Then what?

FLOW TIME
ORGANIC WASTE
MIXED WASTE

15. **PLANTS** have the advantage over other material ornamental objects that recycling is not an issue. Their contribution to climate change is more likely to stem from transporting them over long distances. The best birthday present is a plant from your own balcony. They need looking after and if this doesn't happen, they go through a stage of suspended rejection, right until their looks no longer positively contribute to the surrounding space. The final step is a return to the earth that fed them.

FLOW TIME
ORGANIC WASTE
0

柑 老白茶 陈韵贡眉 白

A beautifully packed and designed Chinese teacake: a simple protective layer of paper made to protect, identify, inform and seduce.

Seduction: a 100% recyclable luxury fashion packaging for lifestyle customers.

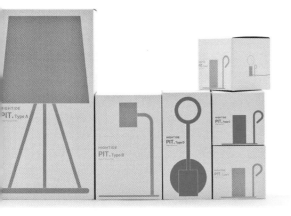

Easy identifiable down-to-earth furniture packaging.

Cardboard shipping package for a shaving kit subscription.

Mesmerising phenomenon: unpacking videos.

Surprising solution for wrapping Japanese jeans.

SHELF AWARENESS

The use of wrapping paper is first documented in ancient China, where paper was invented in the 2nd century BC. In the Southern Song dynasty, monetary gifts were wrapped in paper, forming an envelope known as a chih pao. The wrapped gifts were distributed by the Chinese court to government officials.

The earliest recorded use of paper for actual packaging dates back to 1035, when a Persian traveller visiting markets in Cairo noted that vegetables, spices and hardware were wrapped in paper for the customers after they were sold.

Packaging serves more purposes than one would expect:

PROTECTION
Packaging is used to keep a product safe from external influences, varying from temperature, to moisture, UV light and mechanical forces to human tampering. If you want to sell fruit juice, you just can't hand it over to customers. It should be contained in something.

DOSAGE
If you want to sell a certain amount of something, you need to create packaging to make sure that the right portion is provided. This is true for both the buyer and the seller.

THEFT PREVENTION
Displaying and selling small relatively valuable items, such as jewellery, or memory devices, packaging can serve to 'enlarge' products to prevent petty thieves from simply putting them in their pockets.

BRAND VISIBILITY
Obviously you provide the best product in your category and you want your customers to remember that.

DECORATION
Packaging has to look good and express the promise of what you will find inside.

THE UNPACKAGING EXPERIENCE
The YouTube phenomenon of Unboxing Videos includes videos of people unpacking everything from gadgets and food, to beauty products and luxury clothes. As a matter of fact, one in five consumers claim they have watched one or two. While watching someone else opening the box, we can dream away, opening the package and enjoying it ourselves.

3

cash flows

Fully automated robot driven Autostore Systems distribute all sorts of products at lightening speed while hughe scale outsourced manual labour in the Asian garment industries chases the 'just in time delivery' deadlines for our cheap fashion industries. After all these well organised highly skilled activities all end rather disappointingly in piles of waste on a landfill.

All materials and parts neatly brought in, prepared and arranged, everything in precisely the right order, on the right spot, just in time, quickly assembled with ingeniously designed details, perfect transportation, maximum control until delivery: that is what well-organised production should look like, including of course decent labour conditions.

BUSINESS OPPORTUNITIES FOR CIRCULAR FLOW

In the material sense, producing, selling and logistics that are currently running smoothly. The process from design to shop and direct delivery, has never been so well arranged on such a massive scale and it keeps speeding up. Typically, in fashion chains like H&M and Zara it takes just a few weeks. For food the principle of following seasons (time) and growing vegetables and fruits locally (place), is being supressed to the point that it is vanishing. Fresh produce every day is the benchmark, even if it has to be flown in from half way across the globe. Another category, cheap novelties for football matches and royal weddings, can appear out of the blue in just a matter of days.

The contrast with the part of flow that comes after the production and delivery deadline, the value drop to waste sequence, couldn't be more striking. Lack of responsibility particularly shows, when we consider the groups of low-value products that are mostly addressed in these chapters: food, packaging, disposables, fashion products and gadgets & giftware. With few exceptions, some of which are shown elsewhere in this book, attempts to arrive at a certain degree of limiting and controlling waste, have so far been insufficient. Some companies do their very

best. There are taxes and rules. There are mandatory deposits. This helps, but there is money to be saved and made here. And, as opposed to sitting back and doing nothing, taking responsibility to contain waste provides liveability and bolsters reputations.

When we analyse what happens to value in flow products, it is logical to make a distinction between two possible directions in which improvements can be made.

Slow-down is to direct the attention towards cultivating and sustaining product value rather than allowing it to drop so terribly quickly. This simply entails slowing down flow and turning the partaking products into the kind that is more susceptible to control: good old products that last.

Organising loops is to concentrate on flows and try to organise them in such a way that the harm they inevitably do is brought to a minimum. Slimming down flows and optimising transportation and processing have the potential to save costs, whereas organising flow and segments of it can become a solid basis for new services.

Bodum reusable drinking cups. In line with their motto – make taste, not waste – they're looking at everything from production to packaging to ensure that we don't create unnecessary waste. After all, design isn't merely a product statement.

1. SLOW DOWN

Business models to make products last are of course already known. They address product quality and its long-term sustenance to ensure usability. The models are all based on intensifying use and application. Ideally, replacement of a product that has lasted happens after the optimum lifespan has passed, or when renewal has become unavoidable. This principle may enforce a decrease in the need for new products and thereby reduce flow.

In *Products that Last* the following five business models have been extensively discussed and illustrated. Here we will briefly repeat them, with different examples that fit the flow approach.

Classic Long Life

One could say that this is the godfather of business models for long-lasting products. These are designed to be enduring and that is exactly what they do, no strings attached. Or maybe just one: these products come with a carefully developed reputation that nicely fits the prerequisite of being sustainable. As a matter of fact, lasting products are gradually being reintroduced. Several companies are even moving away from fast value droppers, to replace them with sturdy lasting ones.

Or maybe it is not about sturdiness. A quite intriguing example was made by ceramic design duo JKN (Jan Broekstra and Sander Luske) in the 1990s.

They organised parties and produced copies of banal plastic cups and funnels in biscuit porcelain, as fragile drinking objects. The plan was that guests would smash them after use. Nobody did. The objects just felt too precious to break.

It can of course be that replacing a disposable with a lasting product isn't necessary, since the product at hand already has got the required stamina. It is just value that can't keep up and consumers behave accordingly. Then the business approach needs to be adjusted, which is the core issue anyway. This is for instance the case with fast fashion products. Even items with a cheap reputation often have a quality that matches products of high-end brands. Actually, wearing and using them regularly for a much longer time could be promoted, in line with a slower business concept. This would imply an altogether different proposition. With a lot of effort some brands succeed in promoting an image of forever wearable clothes, but a change of business model from fast and cheap to high quality and slow so far has been too much of a challenge. It is very difficult to replace irresponsible cheapness by attractive slowness. 'Value' is almost exclusively limited to the shopping experience. Regulations could help, but they would have to be enforced internationally.

CONTINUED USE CUPS

Bodum is one of the most reliable brands when it comes to reusable coffee cups. While this stainless steel one isn't one hundred per cent leak-proof, it's

Hellofashion is a continuous collection of 'moniquevanheist-classics', set up to challenge the fashion system. This over time expanding collection of garments, accessories and lifestyle products like furniture and recipes is available for unlimited time.

certainly secure enough to keep most of the liquid inside, when knocked over on your desk. The rubber grip – which comes in a range of different colours – means you can easily drink from it without a handle, and the mouthpiece is large enough for a good amount of coffee with each sip. Capacity is 350 millilitres, which is more than an average-sized mug's worth.

CONTINUED USED FASHION

Whereas prestigious brands such as Prada and Chanel have originally marketed Classic Long- Life items without any conscious sustainability intention, hello*fashion* does just that. Its initiator, Dutch fashion designer Monique van Heist, disagrees with the convention of bringing out new collections every season. Instead she continuously works on one collection that slowly expands. All items remain available. Her collection consists of clothes for men and women, and straightforward jewellery. Interestingly the collection shows that Classic Long-Life products are supposed to be 'circular' despite their withdrawal from the idea of a circular flow. In the case of hello*fashion* there is an exception: a beautiful one-day broche, a white paper copy of a tree leaf.

Hybrid model

Some products can be divided up into a long-lasting part, and an exchangeable component to complete their functionality. The latter is usually disposable. The principle of this model is that, for example, toner cartridges, coffee pads, or detergent containers bring in most of the money. One can therefore state that the hybrid business model is identical for products that last and for those that flow: it always involves a part that is supposed to last that requires disposable parts.

Still, there are flows of products that offer new opportunities when they are viewed from the hybrid business perspective. This could imply that some components that flow could become components that last. Packaging, for instance, could consist of two kinds of components, both of which are supposed to be discarded. One of those parts could be designed to last. The practical example is quite obvious, because it has been around for ages: crates and bottles. They function together and compared to bottles, crates last quite long.

In a wider sense, packaging needs contents to function, which leads directly to the idea of long-lasting packaging. It wouldn't work in every instance, but on a huge scale the sea container is an example of an enduring hybrid product, with all that it may contain as its counterpart.

The giftware category may also benefit from the hybrid view: flowers and vases. It is already possible to subscribe to flowers being sent every week and vases can literally be useful for ages. If plants are grown in a sustainable way in an area close to the customer and showcased in a lasting masterpiece, we have a hybrid business model, partly consisting of compostable material.

Refillable and recyclable toner cartridges by HP. Planting Power offers a professional plant maintenance subscription called Rent a Plant. It specialises in solving climate problems within the office walls.

CIRCULAR CARTRIDGES

Many companies work with a copy and printer machine rental system. The contract includes maintenance, which provides an incentive for printer producers to build long-lasting products. Additionally, selling toner cartridges provides the producer with a constant revenue stream.

HP has taken toner production to a next step in circularity. It introduced a closed loop recycling programme in which plastic from HP ink and toner cartridges are recovered. Currently, more than 80 percent of HP ink cartridges and 100% of HP Laser-Jet toner cartridges contain recycled plastics.

HYBRID FLOWERS

Subscription deals for flowers are a booming business. Many companies and networks are joining in. It makes sense to define the atmosphere of your surroundings with renewable 'materials' in a classic long-life vase. The flowers must be grown organically, and they shouldn't have to travel far. 'Start with Dirt' is a small-scale example in Flushing in the Netherlands. The shopkeeper delivers by bicycle. You can order online. There are large flower delivery networks too. Participating shops, which can be anywhere, take care of deliveries. Customers have to check where products of these chains are grown and under which conditions.

Gap exploiters

Turning leftovers into a business is the most occurring model. Recognising opportunities to make money by bridging gaps is just about the essence of entrepreneurship. Gap exploitation consists of different services to keep products going, such as maintenance and repair. Pumping up value of flow products is yet another option. As it is, there is more than enough material in flow that could do with a value boost. Designers nowadays seem to almost exclusively focus on use continuation by salvaging waste products

Instock restaurants rescue food and turn it into delicious meals. 39% of all food waste happens at the producer due to harvest losses, surpluses or a different size or appearance. 5% is wasted in supermarkets, 14% in the catering industry and 42% in households. Seaweed is now appearing on the market, transforming sea leftovers into good food.

and parts and turning them into something useful once again, without scrapping the material. They see potential for *continued use* in anything, from ashes to cabbage and floating plastics. Ideally, there wouldn't be any gaps. Loops would be closed. For this it would help if it were more development in the area of design to support gap exploitation.

SAVE OUR FOOD: OUTDATED TO LAST LONGER

A new kind of restaurant is finding its way to the public. Named Instock, it prepares meals from ingredients that can no longer be sold in Albert Heijn supermarkets (they helped to start the new chain) and a list of other food providers. As explained in the previous chapter, food past the sell by date has no supermarket value left, but it is completely edible, unlike eating ingredients past the use by or expiration date.

NEGLECTED VEGETABLES

Most people have unknowingly consumed them. Algae are already a multimillion dollar industry and in some regions humans eat seaweed with pleasure. They are the gap between what is available and what is normal. As a new opportrity to provide proteins they are now seeping into the market.

Access model

Product rental is getting more and more popular, particularly since AirBNB turned private owners of homes into landlords by discovering new brokerage markets. Renting used to be limited to cars and washing machines. One could also rent power tools, music equipment and kitchen gear and furniture for special events. Expensive and durable were the typical characteristics. Expensive outfits, like tuxedos and evening dresses were temporarily available too. The typical idea is that one doesn't need to own the things that provide their functionality for a limited amount of time. Let a company be the owner that takes responsibility and gets paid for it.

Currently the market for renting is expanding into the field of personal identity. Whereas trivial functionality, which could include representation, used to be virtually the only reason to rent things, now companies are exploring the market for renting out clothes. Apparently, the convenience of not owning clothes is appreciated. You can always get rid of them with the reassuring feeling that they will end up in good hands, be taken care of and used by others. In addition, expensive clothes become affordable.

FASHION LIBRARIES

Several clothes rental companies have already demonstrated their profitability, both offline and online. LENA the fashion library is an Amsterdam-based provider of knitwear. You can choose between different types of subscriptions that earn you points. With these you can borrow items. If you

'Our collection, your wardrobe' is the statement Amsterdam-based Fashion Library Lena uses for those who want to look tip-top, but do not want to buy something that is worn only once or twice.

like something, you can buy it. The idea of a fashion library includes long-lasting clothes, since the principle intensifies wearing them over a longer period of time. Rent the Runway is a fashion library in the US. It started in 2009 with renting a designer piece for special occasions for a four- or eight-day period. Currently they also provide a monthly subscription, which include shipping, dry cleaning and insurance. With an unlimited subscription, customers can keep the items as long as they want. A cheaper deal requires the client to return items within a month.

Rent the Runway has been investing in a very large logistics platform. Each item is returned to a central location, where is inspected, dry cleaned in an industrial-size machine and steam-pressed. The company owns the largest dry-cleaning operation in the U.S.

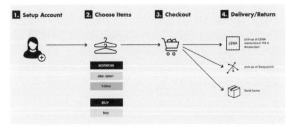

How Lena's Fashion Library works

On their website they say they're in the fashion-technology-engineering-supply-chain-operations-reverse logistics-dry-cleaning-analytics business. This illustrates what attention for flow entails. It is pretty intense.

HOME IN BOXES

Moving provides a small range of examples, since it can be organised in several different ways, each according to a different business model. You can buy cardboard boxes and keep on using them for storage. You can do your own moving here, maybe with a couple of friends. That is the fast flow way, although some people never unpack some of their stuff, maybe until they move to the next place.

The second option is that you rent boxes, often from the moving firm. It is part of their service and it is a matter of providing access. Some movers rent out durable plastic boxes. Movers do most of the moving, with the exception perhaps of expensive crystal ware and other fragile valuables.

The most luxurious way to move is to let the movers handle all of it: packing, transportation and unpacking. Suddenly we have arrived at maximum responsibility outsourcing: the performance model.

Performance model

The principle behind providing a function and getting paid for this, is that the provider, being the owner like in the previous model, is responsible for all the hardware required. The difference with the access model is that hereby users are partly responsible for functionality. The may need to hire a mechanic for repair. In the performance model, company ownership is likely to be an even stronger incentive for deploying long-lasting quality hardware and taking good care of it, to keep costs to a minimum.

Enviropac Rentals is an easy and environmentally friendly way to approach all moving needs.

Known examples are lighting, indoor climate control and printing. The customer pays a monthly fee, and all the rest is taken care of, virtually unnoticed. Fixtures, heaters and copiers are of no concern to users: let the provider worry about them and repair or replace them when necessary.

Some products that flow, which are used to perform a function for a provider, may be made to last longer. This is true for many logistic services. As mentioned earlier (Homes in boxes), a mover may be able to save money by switching from throw-away packaging to reusable crates and boxes.

A different concept could work for sterile disposables for hospitals. Currently a provider curates the process of delivery. All the dirty and unsterile disposable equipment is thrown away, regardless of possible use continuation. The service could be extended to commercial exploitation of the things that could still be useful, by selecting and cleaning them and making them available for customers. Hospital personnel have informally been using superfluous stuff, from gauzes to clamps and scissors. This principle could be formalised. The performance would, partly, get a classic long-life derivate.

Another, different model that can be imagined and which exists in different forms all over the world, is the home decoration makeover service. An organisation offers a deal to redecorate a home once every year for a certain fee. They could replace knickknacks, paintings and small furniture (all exchangeable) from their own stock that they replenish at flea markets and make interesting decorative adjustments to make the house or the apartment feel fresh and adventurous. It could be a way to increase the probability that giftware lasts long enough to be used in several homes. It is not all that different from the service that companies such as 'Lease a Leaf' in the UK provide around plant decorations in offices. Plants may last longer than dead decorations, since they need attention. They have a presence.

One might get help to create an amazing space without the stress or overspending with a subscription for exchangeable decorations, paintings and small furniture.

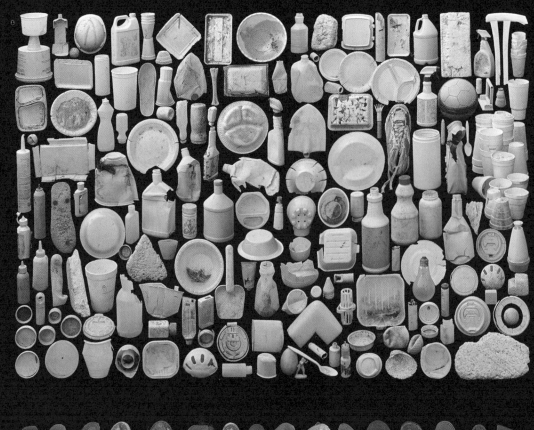

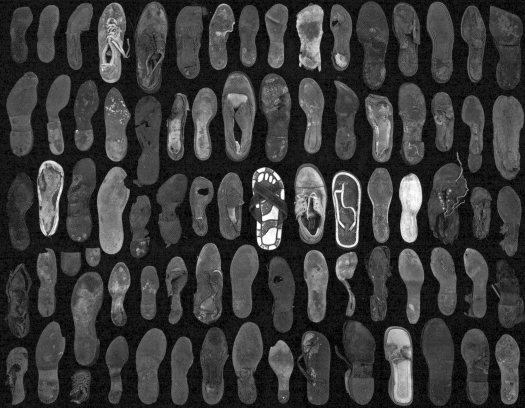

Ivago is the waste management company in Ghent, Belgium. They collect the waste and also manage several recycling centres where you can dispose of larger items or dangerous waste (electrical appliances and paints etc.).

2. ORGANISING LOOPS

Focusing on flow is particularly important, because of the sheer size of it and because of the current lack of control. In addition, attention to flow control helps to discover principles that can also inspire and refine business ideas around products that last, because they are in flow too.

Flow control concerns reducing the side effects of mass flow, rather than the side effects of individual product use. A single plastic bottle of detergent doesn't do any harm, but we're all too familiar with the scale of plastic waste output.

Organising loops consist of a range of activities that have a chronological order and vary in complexity and disciplinary emphasis. Some mostly concern management, others may be more about technology, but still other aspects may be predominantly informational. No matter what, they're all based on cooperation. They're all aimed at reducing the effects that right now nobody feels responsible for. Designers and engineers invent contraptions to clear oceans from plastic, clean the air and store CO_2. Weekends of waste awareness are organised, in which children collect empty bottles and cans in parks and along rivers. Defining responsibilities renders this type of symptom treatment superfluous.

In the current stage of industrial development, flow control is by no means a simple line-up of tasks. It probably never will be. A whole range of different practices exists. There are no organisations that do all of it, but there are companies, successful ones, that handle a sometimes very humble segment of everything that needs to be done.

As said repeatedly, flow implies 'closing the loop'. This expression is used quite often, but its meaning is somewhat unclear. In this book, the *closed-loop recycling system* describes the situation at one end of the 'loop spectrum', in which a specific end-of-life product is separately collected and recycled. Logistical costs may be relatively high, but so is the value of recycled material. Its composition is known, and its properties are predictable. In some instances, recycled materials can be reused in the same product, but that is not essential. The most important thing is that they keep on being exploited to the max.

At the other end of the spectrum we find the *open-loop recycling system*. Here society at large is responsible: citizens and government. In fact, responsibilities are poorly defined. Nobody is certain who is supposed to take care of what. As a consequence of this fuzziness, loads of waste ends up not getting collected. The part that does ends up in waste processing arrangements is subdivided in rather crude categories. Logistical costs are low because of this, but they are still considerable and for most of the leftovers it is difficult to compete with more coherently composed materials that meet certain practical requirements of mechanical, or hygienic, or other nature.

The following activities concern flow organisation and have business potential. The list may not be entirely complete and there can be other tasks that may be in a different order, but one list can lead to another. The point is that system change, or system replacement, requires well-planned implementation.

Artist Barry Rosenthal's collects previously discarded junk along the coastal areas of New York Harbor for his project 'Found in Nature'.

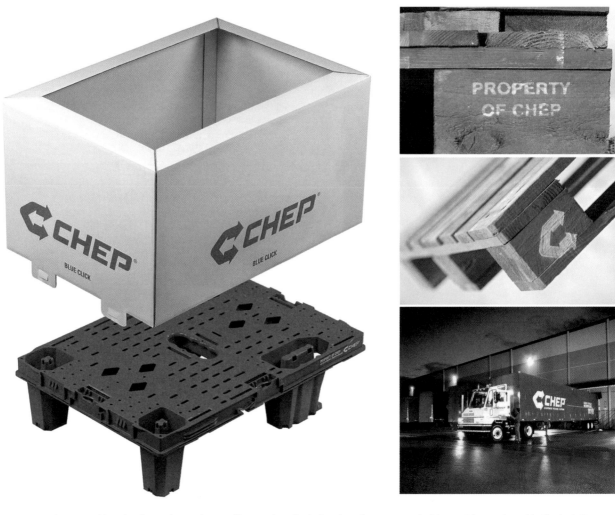

CHEP is a large provider of pallet and container pooling services for industries. They operate in 50 countries and provide the logistic sector with more than 238 million pallets, 14 million containers and 170 million RPCs (Reusable Plastic Containers). Traditionally most pallets and industrial packaging are made from wood, but (recycled) plastic pallets are gaining market share in pooling systems.

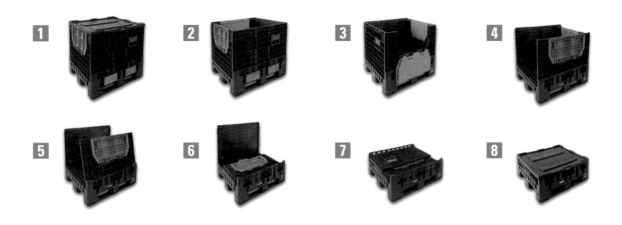

Glass bottles are made from relatively innocuous raw materials and are completely recyclable. Their bulky size and transportation footprint are their downfall. Plastic has a smaller carbon footprint when it comes to transportation, but it's tough to ignore the giant carbon footprint when it ends up in a landfill and becomes a huge pollutant in our environment.

BUSINESS OPPORTUNITIES
① # FLOW MANAGEMENT

- Complete flow management: overview, analysis and control of a particular material flow, or range of flows, from a particular company or network. This implies continuous traceability of all input and output.
- Cooperation: with few exceptions, making sure that rivers of waste remain between the banks requires that stakeholders work together, from resources and energy providers to garbage collectors, government institutions, politicians, civil servants and citizens. Public private partnership can be a particular field of interest.
- An important part of this is slimming down flows to diminish dependence on resources and energy consumption with it. The leanest possible outcome of any flow is total replacement by an immaterial business model, which you will find in the gaming industry and social media.

POOLING SYSTEMS
Most large applications of continued use concepts can be found in the Business to Business market for logistic transport packaging systems. In order to successfully run such a business, it is important to have a profitable scheme, with well-performing packaging products. This depends on smart supply chain control with good relational and operational management.

COINS FOR BOTTLES
The Dutch deposit system for large PET bottles nudges customers to return their emptied containers. It is an almost closed system – 95 percent is returned - and the recycler can trust the material to the extent that without extensive purification he can produce food grade PET material from it. The disadvantage of this system is that not-returned bottles require no processing and therefore contribute to producer income, which is a wrong incentive. Many other European countries and American states run deposit schemes. The German scheme is one of the most successful, with a return rate of 98.5 percent. Manufacturers pay for this system by taxation.

CUP2PAPER

+28%

CARDBOARD / PLA CUP, RECYLING

+104%

CARDBOARD / PLA CUP, FERMENTATION

+113%

CARDBOARD / PLA CUP, RESIDUAL WASTE

+118%

CARDBOARD / PE CUP, RESIDUAL WASTE

+164%

POLYSTYRENE CUP, RESIDUAL WASTE

Total environmental impact in relation to cup2paper (Source LCA Quickscan, Partners for Innovation)

Cup2paper is the most sustainable integrated solution for disposable cups. The cardboard cups have an organic interior (PLA coating) and are 100% recyclable. New Cup2paper cups are made from the recycled paper fibers. The CO_2 emissions that are released during the production and transport of the cups are completely neutralised. This makes the cups 100% CO_2 neutral. SUEZ maintains complete control of the chain, including the collection and processing of the return flow.

By rewarding people for separating their waste, less waste will end up in the subways and streets of Rotterdam, which reduces cleaning costs. Clearing a bottle from the street costs about € 1 while in the water it rises to € 10 per bottle. The ecoeuro machine ensures that the waste is separated and recycled. And a clean public space also contributes to a safe feeling.

BUSINESS OPPORTUNITIES
② REVERSED LOGISTICS

- So far, logistic procedures have been aimed at getting products out, on the marketplace, as mentioned in the beginning of this chapter. It should be just as easy for products to find their way back to their place of birth after use. The appropriate term is reverse logistics. It is already used for repair, maintenance and return upon failure. The aims include further processing, biological as well as technical. We need more of that.
- Traceability: understanding the overview and collecting and analysing logistic data can be a commercial specialty.
- Collecting waste is not equally well arranged in different regions, but even in the best organised countries there is a lot to be gained.

CUP2PAPER
Together with degradable disposable producer Bioodi, material processer SUEZ developed Cup2pape: a controlled flow disposable cup. It's made from cardboard with a bio-based plastic coating (PLA). After use SUEZ collects the material and transports it to WEPA, a recycling company that turns it into toilet paper and tissues. With this concept SUEZ has changed from a waste processor into a 'circular chain director'. They can continuously monitor waste flow quality and processing. Possible negative by-effects are prevented. SUEZ keeps complete chain control.

REVERSE SLOT MACHINES
The subway of Rotterdam (the Netherlands) features reverse slot machines: throw in an empty bottle or can and you get an ecoeuro token in return. It is a discount voucher, to get coffee, or pizza or flowers in local shops, or to contribute to the local zoo. In this way public transport service RET actively increases the recycling ratio and enhances the travel experience.

Ragpickers as still seen in today's less-developed economies used to be common practice in the West up until the rise of the consumer society and the immense increase in prosperity from the sixties onwards.

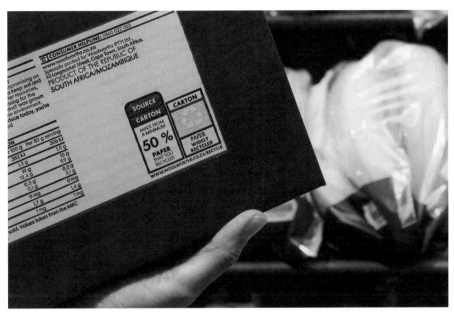

Many people throw out items that can be recycled because they don't realise that those items are recyclable.

BUSINESS OPPORTUNITIES
3 # SEPARATING & SORTING SERVICES

- Collecting waste is not equally well arranged in different regions, but even in the best organised countries there is a lot to be gained. The same is true for selecting different kinds of waste. Technology cannot handle the whole process. The waste coming in can also serve as feedback for the selection process.
- Innovation: change and improvement don't happen by themselves. Particularly in the area of material design there's quite a lot of work to be done. In addition, more emphasis on transportation and processing implies development of new 'flow friendly' concepts.

RAGPICKERS
In many less-developed economies people make a bit of money by ragpicking, selecting waste, even sitting on top of the garbage in the back of moving waste trucks. They often live in landfills, working as the ultimate level gap exploiters.

EPR SYSTEMS
Many European countries embrace separate collection of glass, paper, textiles and plastics. In Belgium, for example, Fort Plus recycles almost 90 percent of all packaging, rendering this country a waste-recycling frontrunner. Their system is a legislative obligation: packaging producers are obliged to take back what they've produced. A similar system exists in Germany. Households get two bins each. An organisation called DSD takes in all the waste and keeps contracts with recyclers.

TerraCycle is an innovative recycling company located in Trenton, New Jersey and it is the international leader in recycling the unrecyclable. The business contin-uses and recycles waste instead of incinerating it or depositing it into a landfill. TerraCycle has begun sourcing what they call "storied plastics," which are post-consumer recycled plastics collected by TerraCycle that have a "story of origin" behind them.

In the offices at TerraCycle everything from wall to floor was crafted with the imaginative contin-use of materials, highlighting the creativity and innovative thinking. The world's first 100% recyclable shampoo bottle made with beach plastic. This bottle is currently sold in Europe and has been so successful that P&G plans on introducing 25% recycled plastic across 500 million bottles sold yearly on its hair care brands.

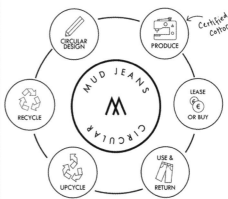

MUD jeans started with the production of jeans with organic cotton in Fair Trade factories. The next step was to get rid of the labels by printing with ecological ink, so that the jeans could be recycled more easily. Now, with the possibility of leasing a pair of jeans, the circle is complete. The jeans remain the company's property, but you can borrow them for a year and then return or exchange them in between for another model. For a 20 Euros entry fee you can already lease a pair of jeans, after which you pay € 5.95 per month for 12 months. In the meantime, you can exchange a model for 10 euros.

BUSINESS OPPORTUNITIES

 # RECYCLING SERVICES

- Regenerating materials to their original quality strongly depends on all processing that precedes it. Organizing the right conditions for recycling and combining material flows completes the flow.
- Retrieving renewable materials from the flow serves two purposes: either purifying the flow to optimize their recycling potential or rendering them suitable for composting.
- Waste from the food chain and other organic waste, can sometimes be digested and composted or be turned into biofuel. Butanol, for instance, can be made from by-products of whisky processing.
- Thermal recycling or burning waste is a last resort. It always produces CO2, of which we have more than we need.

TERRACYCLE

Cigarette butts, coffee capsules, food wrappers, light bulbs – you name the unrecyclable and TerraCycle collects and repurposes it to be applied in consumer and industrial applications.
CEO Tom Szaky founded his organization in 2001 as a student at Princeton University. Working with individuals as well as municipalities, manufacturers and corporations across 21 countries, the company has recycled billions of units of waste to date.
TerraCycle offers a number of recycling options, from free programs funded by companies, to programmes you have to pay for, like Zero Waste Box. These programmes allows recyclers to choose from over 100 flow options for particular kinds of waste, such as disposable plastic gloves, or coffee capsules. The other option is the all-encompassing All-In-One Zero Waste Box. It accepts a wide range of waste with the exception of food or hazardous materials. The process is simple enough: choose a kind of waste to recycle, purchase the appropriate box type, collect the waste and send it to TerraCycle. Shipping is included in the deal.

MUD JEANS

Jeans are the most popular clothing item ever. The brand MUD is an example that may work for other fashionable products as well. The customer can lease a preferred pair for a fixed period. After that she or he can decide to keep it, or to return it. If brought back the trousers can either be turned into a vintage pair, or recycled into something else, like a bag, or a jumper. Interestingly, it is possible to control the flow of a fashion product between company (lease) and customer (owner) responsibility.

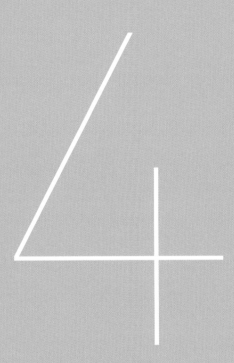

4

exploring the unknown

Whether done by drone, donkey or bicycle: all these flexible and efficient forms of moving products that flow over shorter distances within urban areas remind us of the importance of thinking about designing products with restrictions on size and weight in mind.

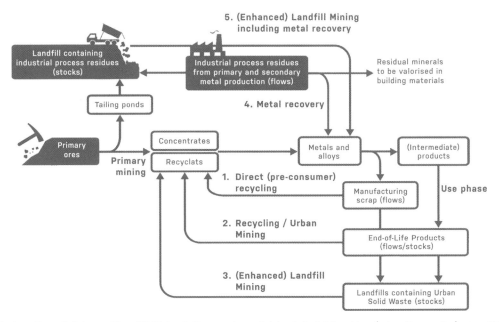

Closing the loop through Enhanced Landfill Mining tailing ponds containing industrial residues/extractive waste (source KU Leuven)

Many designers like to think about themselves as specialists who understand how to discover the functional essence of the product they are creating. They develop it through the eyes of 'the user'. So much has this become a habit that other contexts, such as transportation, trading on the second-hand market, or sequential episodes of processing, are often overlooked.

DESIGN FOR FLOW

Often, products do not possess the resilience they need, nor do they exist in a context that keeps on supporting them with maintenance and care. It is, however, just as true for the products that are in a state of flux, on the move from one processing incident to the next.

It is not all that radical to claim that the ability to flow is simply yet another aspect of functionality that comes with its own set of requirements. Still, it concerns facets that usually remain undiscovered until the products show up as waste. Sure, there are packaging designers, food processing developers, and designers specialised in disposables, fashion, gifts and gadgets. Nevertheless, properties required for the ability of products to pre-emptively spare the environment after consumption or use, are not sufficiently addressed. It is clear that a cheap fashion item will end up as garbage. Yet, prevention is hardly considered. There is little concern beyond the product launch deadline, after which in fact life begins.

This implies concern for everything that supports their value, from maintenance to remanufacturing and communication, all of which is necessary to get the most out of them.

Environmental considerations around products that flow, however, follow an entirely different path, which is to minimise the effects that their transportation and processing have on the environment.

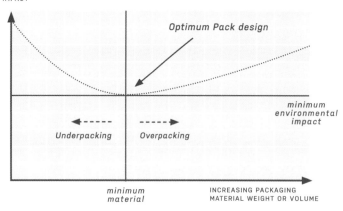

Optimum Pack Design model and the main parameters to be aplied in designs for products that flow

SLOW DOWN: DESIGN STRATEGIES FOR PRODUCTS THAT LAST

Different aims of design for products that last and for those that flow do not exclude mutual influences. One obvious, but not always adequate, strategy to reduce environmental impact, is to turn a flow product into one that is likely to last. We have chosen the new word *continue* for this. This will of course bring about the appropriate design strategies that are discussed in the orange *Products that Last* book. Very briefly it says design can be for:

- Attachment and trust, having all the qualities that make it needed and wanted.
- Durability, with properties of soundness and the ability to endure long-term use
- Standardisation and compatibility, to be able to exchange elements
- Ease of maintenance and repair, implying accessibility of all parts
- Upgradeability and adaptability, to improve functionality in time
- Dis- and reassembly, for access and transportation.

To a certain degree they are interdependent. The ability to dis- and reassemble a product, for example, may cater for ease of maintenance and repair as well.

Depending on flow characteristics, particularly the time during which a product is 'on the road', some of these strategies apply to products that are not supposed to last. Packaging, for instance, may have to be durable when it is doing its protective job. Trust, another example, can be quite important for professional disposables, when it concerns not so much one particular surgical instrument, but rather a certain type, or brand. These strategies work for products that flow, 'while they last'. The ability to disassemble a product can typically be a requirement for nice and easy flow. Reassembly, on the other hand, is pointless for a product that is not supposed to last. Moreover, disassembly in the case of products that flow, reaches much deeper into material composition, as will be described later on, like processing laminated foils.

Products that last also flow. Their episode of use is much more extended, but before and after that, they require transportation and processing too. Therefore, looking from the viewpoint of flow, design strategies to enhance a given product's capacity to minimise environmental effects can also contribute to the sustainability of long-lasting products.

DESIGN STRATEGIES FOR PRODUCTS THAT FLOW

The title of this paragraph could be a convenient way to define the generally right design strategy, but it isn't very concrete. It is wiser to develop a subdivision of strategies that render it easier to focus on certain product properties. Minimum energy transportation and processing, combined with environmental consequences and manageability of the processing output, define criteria by which improvements can be determined.

DESIGN STRATEGIES
① RETHINK

The most drastic change to arrive at a reduction of negative environmental effects is jumping to a different system altogether. Considering a complete turnaround is known as 'rethinking'. There are quite a few examples in this book.

Fast-moving consumer goods do need to be fast moving in all respects. If the content of the package is eaten or used up presto, this does not mean packaging has to lose all its value simultaneously. Alternative delivery methods, including even a package- free system, seem farfetched. Nevertheless, they're an option. Here is a line-up of some examples in which our current system has undergone serious rethinking.

Retail is showing progress in several cities in Europe. In Berlin the shop 'Original Unverpackt' is open since September 2014 and sells over 600 products. Another shop in Antwerp, called 'BeRobuust', is an organic product store that sells according to a similar package free concept. Shoppers are required to bring their own pots, bottles and bags and they can fill them up from large containers in the supermarket. This does not imply that no packaging is used at all. The bulk goods still enter the store in packaging.

These successful examples show that losing the packaging can be a viable business strategy. Some though, are failures. The package-free supermarkets in Groningen and Utrecht (the Netherlands) had to close soon after the opening.

Original Unverpackt first opened its doors on September 13th, 2014. Ever since, people have been flocking there to buy their weekly groceries and household supplies. With a product range extending from groceries to spirits, cosmetics and books, it is simply a supermarket – except that there is no disposable plastic packaging.

At vegetable company Eosta / Nature & More they believe that organic products should not be packed, unless there is no other option. First and foremost they want to get rid of plastic. Their quest has taken them from PLA and other non-oil-based plastics to sugar cane material, compostable stickers, carton trays and, Natural Branding. Natural branding concerns creating an image directly on peels. A high definition laser removes pigment from the outer layer of the skin of the product. There is no detrimental effect on taste, aroma, shelf-life and edibility. Just with avocados, Eosta will save 750,000 plastic flow packs in 2017.

The company Soap by Mail had to overcome several barriers. They needed the help of a chemist to develop concentrated soap. For packing they used Polyvinyl alcohol, a dissolvable material. This proved to be extremely useful, since it actually improves the detergent. The box had to be designed to fit through a letterbox defined as 'large letter' by the UK's Royal Mail.

In some cases, the product can be changed to render packaging superfluous. Soda stream promotes itself with such a feature. They don't provide readymade soda mixes. Rather than shipping bottles around the world, they provide a machine with a carbonator bottle. Depending on its size, it makes from 30 up to 130 litres. Lush, a personal hygiene company, offers a broad range of products that come as solids, where liquids would be expected. For instance, they sell shampoo bars and hard deodorants, thereby altering the packaging concept. Some products do not require packaging.

Angus Grahame set up Splosh in 2012 with the idea that there must be an opportunity to sell household cleaning products online. However, the typical size and weight of the products was unpractical. Therefore, he embarked on a total makeover. Splosh customers now purchase a starter box containing a range of reusable bottles. In each of those, consumers can make their own cleaning products by mixing a sachet of concentrated liquid with warm tap water. Sachets in packages arrive through the mailbox. Several companies offer a similar alternative to liquid washing detergents.

For a lot of liquid consumables, a refilling system can be a solution. Concentrating the juice or the detergent can save a lot of packaging and logistics. For the customer this may require some extra handling. Good examples of this option are sauce dispensers in restaurants and soap dispensers in hotel bathrooms. They may save money but do require refilling and cleaning.

Many food products are already packed in their own skin. That is what contains the food and protects it, and in the case of bananas, provides peeling convenience. There happens to be a rule in the Netherlands that organic products have to be clearly distinguishable from the nonorganic ones. This is usually achieved by packing the organic ones in plastic, which is cheaper than the other way around, because there's simply fewer of them. This is awkward, particularly from the viewpoint of organic customers.

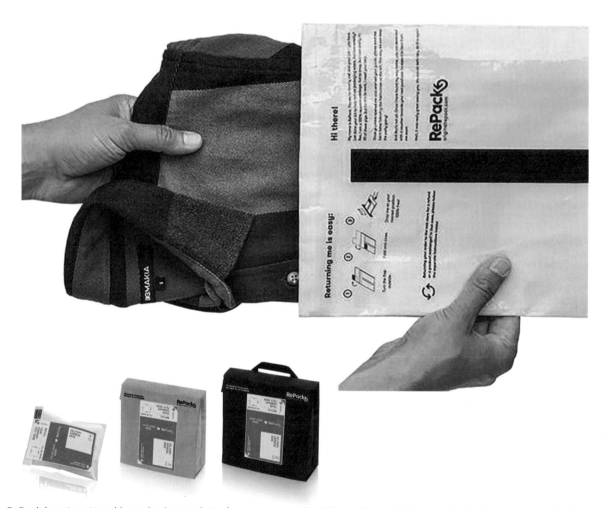

RePack boasts returnable packaging products for e-commerce: 'It will save the world from trash. It brings people and retailers together in a loop of good. A super simple solution to an ever-growing problem'. It is a closed-loop system of returnable packaging in three adjustable sizes. When emptied, packaging can be flattened and dropped in the mailbox to be returned, free of charge. There is collaboration with brands, such as Filippa K, with which you can opt for Repack for an extra 4 euros. Once Repack packaging is returned, the client is rewarded with a voucher to be spent in the web shop of the same brand.

DESIGN STRATEGIES
2 DESIGN FOR CONTIN-USE
(FORMERLY KNOWN AS REUSE)

Humans aching for convenience cause the market for disposables to be continuously growing. In some cases, the increase can be halted through clever alternatives. Where value tends to dissipate very quickly, redesign for a continued use concept should always be considered. Lots of examples are already quite common. Think of pool pallets, crates and other durable packaging systems.

The growth of e-commerce particularly increases the use of cardboard boxes (sometimes to the extent that they pack one sheet of printed paper in a box that won't fit through a letterbox).

For the consumer market it is more difficult to organise return logistics then for businesses. Logistics involves more parties: producers, wholesale, retailers, consumers and transportation services. Yet, it's possible: reuse options are beginning to show up.

Glass has a different potential. Since the 1980s the Dutch have become accustomed to a deposit system for glass beer bottles. After return, bottles are cleaned and refilled, which is estimated to be eight times more environmentally friendly than single-use bottles. In France, on the other hand, a number of supermarkets have introduced a wine refill machine, stimulating customers to reuse wine bottles.

Glass happens to be very suitable for reuse in packaging systems and no matter how old and familiar it is as a packaging material, it can still be applied in new ways.

Clothing reuse options are more common and new opportunities emerge continuously. Many textile recycling schemes, for example, are financially feasible because of the high quality of pieces that are brought to collection boxes to be resold. Clothes swap parties stimulate contin-use too. Design for reuse is a matter of focus on high quality, adaptability to different sizes and a certain generosity in style: an item shouldn't devour too much attention at first sight.

GO Box is an initiative in Portland, United States, that provides a reusable lunchbox. You can order your food online with collaborating restaurants and make sure you get it served in a GO Box. Afterwards, you return the GO Box at one of the drop-off points close to a collaborating restaurant. Around 30 restaurants in downtown Portland have joined in. Membership amounts to 18 dollars per year.

Beautiful scuffs: To reduce Heineken's carbon footprint, it would be best to switch from one-way to returnable and refillable bottles. Those, however, are considered less classy. They scuff, get damaged and clients perceive them as less appealing. The brief for design studio NPK, was to design a bottle that becomes more attractive with ageing.

The returnable bottle is re-named to get rid of the backward looking old-fashioned idea of returnable bottles. It's called the FOBO: forwardable bottle. The shape is completely new, without changing the category. It still typically looks like a container for lager and not water, juice or olive oil. The hero is not a paper label, but a large embossment. This is likely to get scuffed over time. The more it scuffs, the whiter it becomes, the more it will stand out and the more attractive it gets. The concept is being tested in France. In the white screen-printed area each bottle

has a unique etched code. Via fobo.fr consumers can see where that particular bottle has been, what it went through, which special cool night, or which surprising encounters it witnessed. You can add your own message to forward to the next consumer, as a chapter of the story chains. The more each FOBO is being forwarded, the more legendary it becomes. Over time all FOBOs will be connected to chains of true consumer stories and will be unique and authentic through those. Simultaneously, FOBOs may reduce Heineken's carbon footprint.

Filippa K is a Swedish fashion brand that ambitiously aims to produce simple high-quality products for long-term use. They have a renting service for clothing in their general stores and they collaborate with a local second-hand business in Stockholm, where they founded a shop for their second-hand clothes. People can bring those to the store and when they're sold they get 50 percent. Unsold items are returned to the owner or donated to charity.

Swedish high-end fashion company, Filippa K, has a long-term goal that by 2030 its entire collection will be designed and produced following the principles of circular economy – reduce, repair, reuse and recycle. The brand is already offering products and services on the way to this goal.

For Filippa K, quality and design for a long product life are of utmost importance. These are qualities that can't be compromised since a long life maximises environmental benefit.

As a step towards a circular economy, you are welcome to return old Filippa K garments that you no longer use. These will either be sold in the Filippa K second hand store or given to a selected humanitarian organisation. Return a Filippa K garment to your local store and receive a 15% voucher to be used on your next purchase. In Sweden, we also work with Cirqle, an app that guides individuals to stores that collect used clothes. Once the user has donated clothes to a Cirqle connected store, he or she gets to choose a reward through the app, e.g. a digital discount.

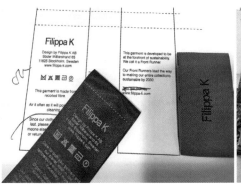

Among other activities the company Lightweight Containers exploits their KeyKeg. It is a pure PET container for beer, wine and other fluids, that is designed to replace the metal ones. They are 75 to 85 percent lighter than their predecessors, because PET has a lower density, but also because it allows for a much smaller wall thickness. The main components are the vessel itself and the bag inside that contains the fluid. It can be pressurised with air in between the vessel and the bag to press out the contents. Vessels are almost completely recyclable into new ones. The bag (about 15 percent of the whole) is laminated and can be recycled on a lower level.

DESIGN STRATEGIES
3 DESIGN TO SLIM DOWN

An obvious, but often overlooked, design strategy concerns 'slimming down' the product, or reducing its weight. Apart from diminishing the mass of material involved this will reduce transportation and energy consumption.

The clearest illustration comes from aviation: if the airline reduces the weight of one airplane by not taking just one 25 grams bag of salted peanuts, this will reduce yearly operational costs by over 2000 dollars. Forget about handing out peanuts altogether and the company saves 800,000 dollars each year, per aircraft.

If weight reduction is simply a matter of diminishing the amount of material involved, it is already standard procedure, since thinner walls and lighter materials may save costs in the supply chain. Indeed, PET bottles are losing weight. Research from the IBWA (International Bottled Water Association) found that the average gram weight of a single-used water bottle had shrunk by 32.5% over the previous eight years. Corrugated containerboard for packaging is also getting lighter, particularly in developed countries. Similar developments occur in other materials, such as glass and metal.

The UK-based chain Marks & Spencer's, which includes a food section, started the so-called 'Thin Air' project. This involved redesigning 140 of Marks & Spencer's best-selling products, such as crisps and popcorn, by reducing the amount of air in the top of the bag: same quantity of snacks, requiring less packaging materials.

Popcorn bags were the best achievers, with a 37 percent slimming down in pack size. All of the different changes together have resulted in a reduction of 75 tonnes of packaging per year. According to similar reasoning, Coca Cola has developed a lightweight aluminium lid for its cans, thereby reducing the material use and the CO_2 emissions that go with it.

An important incentive is environmental legislation, such as the European Directive on Packaging and Packaging Waste, which encourages packagers to think lightweight. Consumers are also driving brand owners to go easy on weight and volume. Social media are full of examples of 'unpacking experiences' of ecommerce products with excessive amounts of boxes and layers around them.

Some product managers claim that slimming down packaging any further is a hassle, since the financial benefits of the optimum solution are clear and therefore already pursued. Here the amount of packaging materials is weighed against cost to sufficiently fulfil its protective function. Research has shown that this 'optimum' is often based on personal preferences. More often than not there is space for higher reduction targets, if it were just for the observation that mass doesn't just depend on material thicknesses, but also on other material properties and on functional requirements.

The design process of slimming down should involve consideration of conceptual flow alternatives with different combinations of materials, structural solutions and manufacturing procedures. In exceptional cases, it may even be viable to get rid of the product altogether: think salt peanuts and airplanes.

Weight reduction for the four other categories depends on different considerations. In food it can be a matter of resorting to the essence of digestibility and taste, but the indigestible, such as peels or stems, may still be necessary for conservation. Thinking the other way around, food mass can also be reduced temporarily, by taking out water to put it back in when consumption is at hand. Fruit juices are mostly transported and stored as a concentrated ingredient that is diluted and packaged for selling and drinking. For herbs freeze drying serves to conserve them, but it saves their weight too.

Material reduction is already part of the rationale behind disposables: one wouldn't want to discard too much of materials that have been paid for. Still reconsidering functional essence can lead to the development of a different solution. Sterile scissors to cut suture material could for instance become superfluous if suture thread could simply be broken by hand.

Gifts and gadgets are tools for social lubrication and identity awareness. Generally, the former preferably look precious and/or feel heavy, whereas the latter quite often are small and light versions of familiar function providers: pocket vacuum cleaners, travel irons, those kind of things. The best way to reduce mass here is to try and find immaterial alternatives, but they are charming in their own right and interesting as lightness inspirations.

Actually, we have a link here with products that last. Saving weight generally is not expected to increase the lifespan of products (although it can under certain conditions that will not be discussed here). However, reducing product mass is likely to contribute to a reduction of the environmental burden. Products that last flow as well. Moreover, many products that last are the carriers of flow: transportation machines. They can be considered packaging that is provided with the miraculous gift

of powered carrying. Irony has it, that most of what vehicles move around is their own weight. Even airplanes do that. Electric cars particularly are heavy because of battery weight. Slimming those down directly extends the range, in a much smarter way than by increasing battery volume.

Fast fashion has a different background altogether. There is a general trend for clothes to become lighter and speed is considered important and increasing, particularly since production happens far away from the Internet warehouses and shops that provide the items to customers. The implication is that air cargo companies are competing with price per volume offers to acquire a piece of the fashion cake.

There is a caveat. In general, economic growth tends to undo the effects of mass reduction. That is not an argument to refrain from it, because things would be worse without slimming attempts. Instead it is wiser to start analysing what currently drives economic growth and developing ideas for less-damaging alternatives.

Unilever purchases 2 million tonnes of packaging a year, and in their Sustainable Living Plan they have committed themselves to becoming 'circular' as a zero waste business. As such they have succeeded in decreasing the packaging weight per consumer by 15 percent in six years. There's more. MuCell™ injection moulding, developed by Trexel (Wilmington Massachusetts) is a technology whereby plastic material is turned into a microcell foam without using a foaming agent. Tiny air bubbles reduce the amount of material in the mould, without loss of quality. It has allowed Unilever to reduce the plastic amount in bottles by up to 15 percent.

DESIGN STRATEGIES
④ DESIGN FOR RECYCLABILITY

Technically, everything can be recycled. Milling down something, or melting, or dissolving and separating elements to use them for producing something new, however, is not just a matter of conquering technical challenges. It also involves the identification, selection and subdivision of ingredients and combining them again, procedures that precede what happens in shredders and may continue after that.

With this (sometimes very) complicated scenario we can create a fresh material, but it will have to be affordable when it is compared to its virgin competitor. No matter how easy it seems to regenerate those cheap omnipresent packaging plastics, they can be too expensive to recycle into materials that are pure and cheap enough to compete with the virgin original.

A promotional term has been introduced in previous years, derived from the word recycling, 'upcycling' (as opposed to 'downcycling'), which suggests recycling can increase value. This may sometimes happen, but it is not clear which value is addressed and what happens next. Fleece, for instance, is a fabric that is woven from fibres that are made from recycled PET bottles. Fleece clothes may be more valuable than PET bottles, but it is doubtful whether the PET fibres are better than the PET granulate. Moreover, once it has been worn as a sweater, or whatever, fleece value has evaporated. Currently, there is no next cycle.

Apart from the confusion that the seductive positivity of the word 'upcycling' tends to produce, authentic examples are difficult to find. It is better to think of recycling as a process that, ideally, can be repeated for years on end. Often recycling is already considered a satisfactory procedure if it happens just once and is in fact just a postponement of winding up in a landfill.

PET happens to behave very well in a recycling trajectory, provided contamination is kept to a minimum. Regulations around food safety are strict and demand high quality to prevent chemicals from migrating into the food. Large brands, such as Coca Cola, are introducing recycled fragments in their bottles. In UK bottles the company has increased its contribution of recycled material from 15 to 25 percent in the past decade.

Bottles made of 100 percent recycled PET bottles do exist, but claims are brought forward that this is impossible for all brands and products because properties degrade and supplies of pure recycled PET are insufficient.

Aluminium products have been recycled over and over again for many decades, for the simple reason that producing the metal from its ore bauxite

A brand new Swedish company, called Re:newcell, knows how to recycle cotton. Old clothes are collected and shredded, then turned onto a porridge-like substance. Zippers and buttons are removed. Consequently, the porridge is broken down to the molecular level and turned into fibres to be used to produce thread for weaving. Pure cotton works best, but a mix with other materials is not impossible.

requires twenty times more energy than recycling it. Many companies, such as Apple and Nespresso, boast about their environmentally responsible use of recycled aluminium, but there is nothing unusual about this kind of cost saving.

Recycling paper has reached a high level of control. It is not perfect yet. Consumers are prepared to separate all paper from the rest of their waste, but are often unaware that grease and small amounts of leftovers spoil paper's processing potential. In the world of paper, waste pizza boxes are notorious. There are initiatives increasing the recycling rate. The Good Roll, a toilet paper company that introduced a 100 percent recycled toilet paper, also provides the toilet paper as a service. You can get a home or office subscription.

For prosperous recycling, two things are important. Firstly, the composition of materials needs to be simple and it should be possible to easily separate blends. In the next chapter we will go into more detail about this. Secondly, all materials must be traceable or recognizable, preferably with the ease you can observe in entertaining forensic procedures in crime series on TV. Currently, many materials can already be identified, for instance with near infrared technology. Nevertheless, they are not sufficiently precise when it concerns contamination. Careful

flow control throughout production, transportation, use and processing and contamination prevention is preferable.

Generally, it is wise to involve recyclers in your developments. They know the latest developments and can explain exactly what happens, once they unleash their system on your products.

Applying recycled materials is not so much a matter of a design strategy, but rather a buying policy. Nevertheless, it needs attention, since the demand for recycled materials is still low, whereas interchangeability with virgin materials is increasing. Their quality deserves careful monitoring and markets are starting to appreciate the use of recycled materials. Currently it is an advantageous contributor to a positive brand identity. Apple and Nespresso have already been mentioned, but nowadays most companies take any opportunity to present themselves as 'green'.

As it is, this epithet is mostly used for the inclusion of recycled materials in products. Design for later recyclability is still lagging behind. The term is often misused to state that recycled materials have been used, or that packaging is made of bamboo or other renewable materials, which may be bio-degradable, but that is a different strategy.

In Nepal, two Dutch entrepreneurs, Melvin Loggies and Jasper Gabriëlse, witnessed how women used the empty shells of the Sapindus mukurosi fruit (soap berries or soapnuts) to do their laundry. 'If they can do it in Nepal, we can do it in The Netherlands', they thought. They shipped a big bag full of the shells to Holland and Seepje began its journey. Since Sapindus shell waste is the main ingredient for Seepje detergents, a waste source had to become the main ingredient for their bottle as well. They found this at a supplier involved in recycling post-consumer waste from HDPE milk bottles in the UK. This material, with all its shortcomings defined the packaging design requirements.

DESIGN STRATEGIES
⑤ DESIGN FOR RENEWABILITY

Many materials are far too complicated for technical recycling. As it happens, many of them consist of ingredients that grow and feed on biodegradable materials. Most of them are plant-based. Their main advantage is that breaking them down takes little effort. A bit of shredding may be necessary. The ability to rot renders them almost harmless.

The organisms that break them down produce greenhouse gasses. On the other hand, plants turn those into water and oxygen when they're growing. There are two kinds of renewable materials: those that are bio-based and those that are bio-based and also biodegradable. The former group includes many common materials such as wood and leather, but the word refers to modern materials that have undergone more extensive processing such as bio-plastics and bio-composites.

Biodegradables can be useful in specific circumstances, particularly when degradation contributes to functionality. In agriculture, replacing PE mulch foil with its biodegradable counterpart, saves removal effort. Another interesting application could be biodegradable packaging for food on inflight meals. They can be composted together with the leftovers.

Their relative innocence renders biodegradable materials suitable for strategically replacing technical materials in many applications. Not all products can consist of degradable materials. It may render them too vulnerable, for instance for water. Anyway, the replacing process is evolution in progress. The main difficulty is separation of flows. Confusion and traceability still often result in mutual material flow contamination, leading to hampered recyclability in plastics, and to plastic pollution of composting sites. Another limitation is advanced processing. Complicated high-grade plastic parts can be turned into granules for injection moulding new complicated plastic parts in a jiffy. A complicated wooden part is not as easy to reproduce, for it is not born in a mould that can repeat the same trick.

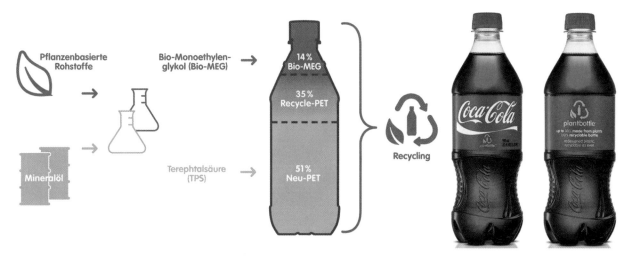

In 2009 Coca Cola introduced PlantBottle™. Not only is it made from bio-based plastics, but it is also fully recyclable in the regular PET recycling scheme. So far, the beverage producer has made tens of billions of PlantBottle™ containers, resulting in a considerable reduction of potential CO2 emissions equivalent to burning a million barrels of oil.

Peeze developed a biodegradable pod as an alternative to existing aluminium and plastic ones.

Bio-materials can be used in all flow products, but they are not equally normal. We have already established that the main drawback is maintaining flow purity. In packaging this can be complicated.

Together with ATI (Advanced Technology Innovations), coffee refinery Peeze developed a biodegradable pod as an alternative to existing aluminium and plastic ones. The new capsules and the sealing film both consist of PLA (polylactic acid), a compostable thermoplastic. It complies with the European standard for compostable packaging EN-13432 and is therefore allowed to wear the Seedling logo. A special grade of PLA needed to be developed

to both withstand the temperature of hot water and allow correct perforation under high pressure. In addition, the material has to retain the coffee flavour.

For disposables the ease with which flow can be controlled depends on context. The loop of food containers as used in coffee and snack restaurants is not as closed as it can be in professional environments, such as hospitals.

For Rijnstate Hospital in Arnhem, food designer Katja Gruijters developed a new disposable tray and dinner service. She worked together with product designer Fons Broess and the product is being produced by

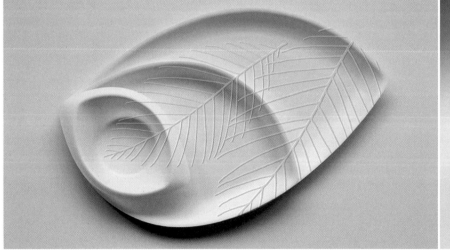

Compostable crockery designed for Rijnstate Hospital in Arnhem by food designer Katja Gruijters and product designer Fond Broess

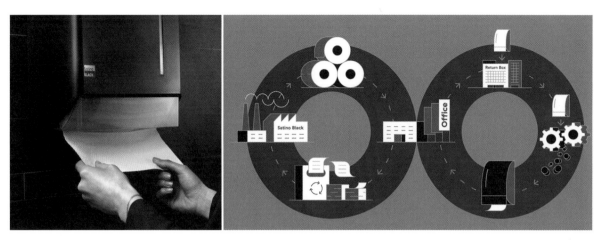

Used office paper, coffee cups and drinking cartons? Dutch company WEPA turns them into Satino Black. You wouldn't think it with such a name, but it is toilet paper. It is a regional setup, so transportation is limited. Its production is certified to be100 percent safe and CO2-neutral. Satino Black hygienic paper has won the international WWF Paper Award.

disposables manufacturer Papstar. The intention to use it for all meals provided the designers with an opportunity to rethink the entire system.

The most notable characteristic is the complete disappearance of the tray, or its merger with the plate: the 'serving plate'. The set consists of three plates, the smallest of which is also the deepest and can contain soup or porridge.

It is designed to become part of an entirely new composting system with considerably simplified logistics. The plates consist of cane fibres, a plant-based material, which is represented in the looks.

After pressure from social media, computer producer DELL has decided to work towards sustainable product packaging. They have for instance reduced the size of their boxes by ten percent. Dell has also started to use wheat straw, which is agricultural waste on which mushrooms are grown: rice hulls or wheat chaff are placed in a mould and injected with mushroom spawn. Five to ten days later, the mushroom roots have created a connecting structure defined by the mould. After drying in a kiln, the straw composite shapes can serve as cushioning in boxes: cheaper and more energy efficient than plastic foam or cardboard.

GREEN PACKAGING: *Dell has been an innovator in using materials from nature for packaging, having already incorporated bamboo and mushrooms. Now, the next evolution in sustainable materials is wheat straw.*

Swiss company Freitag, well-known for the colourful bags that it produces from used truck tarpaulin, has developed a biodegradable fabric, surprisingly named F-ABRIC.

Hemp is one of the fastest growing plants and was one of the first plants to be spun into usable fibre 10,000 years ago. It can be refined into a variety of commercial items including paper, textiles, clothing, biodegradable plastics, paint, insulation, biofuel, food, and animal feed.

Biodegradable fashion items are in the stage of experimentation. There still is a long way to go, even for them to reach a similar level as packaging now. The whole issue of biodegradability in products is gaining momentum.

Swiss company Freitag, well-known for the colourful bags it produces from used truck tarpaulin, has developed a biodegradable fabric, surprisingly named F-ABRIC. The fibres are hemp and flex. Naturally the material is non-toxic. Apart from the fact that the material is grown nearby, with limited use of water and energy, it can also be composted after years of wearing. It was first tested and used as company clothing and now teh company promote their clothes for brain workers.

Many people yet need to be convinced. That is certainly true for gifts and gadgets. It would be wonderful if plants were to become standard gifts. Or food of course, because renewability of food is not an issue. Not really.

Be both grateful and patiënt with a wonderfull plant that can grow a personal harvestable pineapple. Looking for the perfect gift gadget? Don't just buy chocolate, but give your friends the real taste of it.

This design looks deceivingly simple, and that is exactly what makes it so smart and attractive, especially when realising the impact these ideas have when applied to the massive global scale of this brand. Great results both on an economic and envirenmental scale can be achieved when big companies are willing to invest in rethinkers.

Contrary to what these images suggest, this is not an innovation innitiated by the large multinational. It is an initiative from an American student named Bryant Yee who dared to rethink packaging for LED lights. Innovations always come from a roots level, never topdown. Companies know about running their business while it's people who provide the knowledge and ideas.

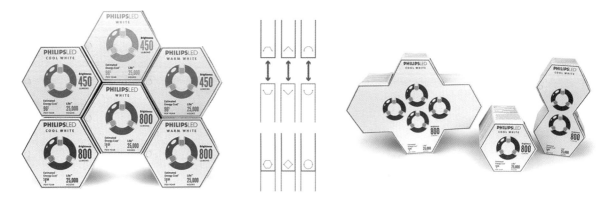

Packaging is not always about industrial scale and advanced technology as this tea packaging by Tin Chan proves. It can also be personal and handmade where playfulness and improvisation lead to solutions that almost automatically seem to meet all requirements.

INTEGRATED FLOW DESIGN

CLEVER LITTLE BAG

Puma introduced a contin-usable shoebox packaging that will reduce the use of paper for shoeboxes by 65% and carbon emissions by 10,000 tons per year. PUMA partnered with designer Yves Béhar, from Fuseproject in San Francisco, to rethink the way the millions of pairs of shoes that it sells each year are packaged. Less packaging implies reduction of the amount of raw materials, water use, and energy consumption for production, shipping and disposal.

In line with the Clever Little Bag Puma also introduced a biodegradable shirt bag, which can be half the size it used to be just by folding the shirt one extra time.

INSIDE AND OUTSIDE THE BOX

This is a project initiated in 2011 by Bryant Yee from the University of Michigan School of Art & Design. It aims to bring awareness to socially responsible LED packaging. It strictly consists of 100% post-consumer recycled paper, promotes recycling old bulbs, and supplies the consumer with up-to-date information through clear graphic design.

The initial prospect was to develop a package capable of returning dead bulbs to the manufacturer for recycling. This resulted in a design that can quickly transform to fit any bulb placed back inside, while protecting it from being crushed during transport. It has a springy, flexible, protective, interior core made of just folded paper and is completely glue-free.

With the appropriate graphic labels, focusing on lumens as opposed to watts and emphasising annual economic savings for the consumer, these changes are key drivers in helping consumers make educated purchasing decisions as they transition to more energy-efficient lamps while contributing to the overall increase in the adoption and promotion of sustainability.

Folding experiments for the LED packaging

5

families that matter

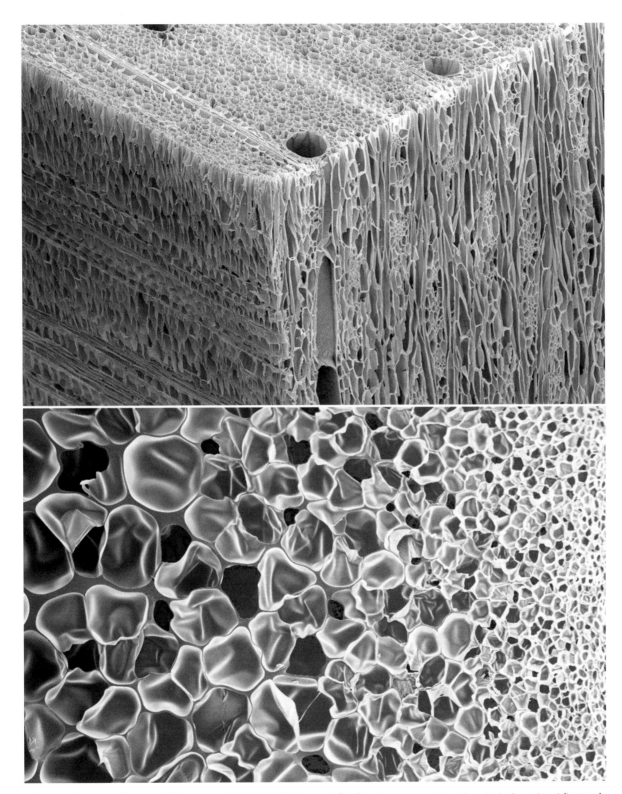

Electron microscope images of the corner of a block of balsa wood (top) and a cross section of a single foam bead (bottom)

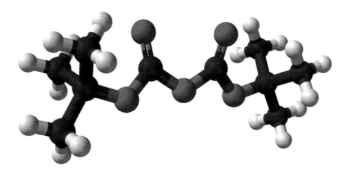

Materials are visible and tangible, also when the products that they have defined are broken, mutilated or shredded. That may be the reason why they get most of the blame when it concerns environmental and climatological issues. Nevertheless, it is us who define material flows. As a consequence, we have to pick, or design, materials in such a way that harm to our living environment from all our cyclical and circular hustle and bustle is minimal. The amount of options is huge. We have to manoeuvre very carefully among these to arrive at the best possible range of treatments.

MATERIAL FLOWS

The word 'material' itself already has different meanings: it can indicate a more or less pure substance, such as bronze, or ABS (Acrylonitrile Butadiene Styrene, to name an often applied structural plastic), but the word 'material' is also used for organised combinations of elements, like wood, which has grown from many cells to become a structure, or like fabric, which consists of woven threads that in their turn have been twined from fibres.

The confusion between materials and material structures has to be made explicit, because properties need to be clear. Foils and fabrics behave differently.

The most annoying question one can ask an experienced circular designer is: which is the most sustainable material? The only legitimate (and just as annoying) answer is: it depends. Firstly, a given piece of 'material' is not some museum piece in a glass case on a pedestal. You can touch it, treat it in various ways, by hand or with machines, or fire, for all kinds of different purposes, from keeping something upright, via guiding heat, water or electricity, all the way to representing an idea or an identity.

A material, however, always operates within a context. At the basic level of technology any given material is always chosen in combination with what you can do with it – processing – and what it is supposed to do. That is the concept. A concept is a set of rules that define a result. This can be a product, or a service. A crude example: cans for beverages are supposed to be light and strong, for transportation purposes. Concrete has lower density

Electron microscope images of Goretex, a lightweight, waterproof fabric for all-weather use.

than aluminium and it is strong indeed. Still, no one considers producing concrete beer cans. That is because the refined shape and the opening tab are far too precise functional features to be made in concrete (process) and because concrete is not very good at being a thin-walled container for carbonised fluids transported in boxes together with many other pressurised thin walled containers. It would most likely break (concept). A concept can be tested to find out if it can become a definitive result, in this case a packaging product. Often various concepts are developed to be able to make decisions on one final design.

Next, we're talking 'sustainable'. This term originally describes an important characteristic of an ideal society and its development, in which the entire world population can live a healthy life, with sufficient nutrition and education. In certain respects, aluminium can contribute, in others polyethylene has a role to play, in yet others plywood helps, and so on and so forth. This however, can never be generalised into some material being

'sustainable', like a material cannot be belligerent, or docile, or industrious. Sustainability is defined by our deeds. Materials are a consequence of design and decision making.

DESIGN PRIORITIES

Materials can be used responsibly, contributing to product functionality with care for the effect they have on life and the environment, depending on their properties, but also on context. For the context of products that flow, it is obviously flow in its entirety that defines context. Of course, there is no standard flow, but largely there are common denominators, as written many times: transportation, processing and control.

In a process of iteration their respective environmental effects can be weighed against each other to discover optimum solutions. This is no simple matter, particularly when the aim is to rethink and maybe even replace the entire flow concept, such as deciding to produce blouses in newly developed paper, to be thrown away with paper trash. There are three levels of design that need to be considered:

A. FLOW DESIGN

Overall flow mainly consists of transportation, processing and control, but depending on the product it can be refined. Storage could be another one, and so could harvesting. The point is that their importance to minimising the harmful effects of flow has to be established.

A practical example is the Keykeg PET beer container. One would think that for such a product metal would be preferable, since a steel container can last, but the logistic advantage that saves considerable amounts of truck fuel easily outweighs the use of a lightweight – non lasting – plastic structure. In addition, its recycling system is getting well under control. For fruit, considering flow could focus on harvesting and ways to minimise the amount of produce that is lost in selection, which, for instance, could be compared to developing the production of a new a new beverage made from misshaped fruit, which may require packaging.

Misfit Juicery produces cold-pressed juices made from 70-80 percent recovered fruits and veggies like carrot sticks and watermelon cubes.

B. PRODUCT DESIGN

Material function and required properties can be defined when the context is more or less set. More or less? Remember, design is a repetitive process: earlier decisions in the flow concept may require reconsidering. Anyway, to be able to flow the product will have to meet requirements, and therefore consist of one or more materials that can help meet those with their properties, characteristics and behaviour.

Some properties are obvious, such as strength, stiffness, the ability to withstand or guide heat, insulation, density, price, solvability in water, etc., the things listed in tables. Other, more concrete properties need to be discovered through testing in practice: trial and error. Finding out if ink will hold, or a plate will fold in the right manner, or if you can laser a logo on mango skin, cannot be done on Google.

Materiability is a research group that attempts to bridge the gap between design and science. They experiment with new materials and technologies, like using phase-change materials to build structures without moulds.

It is crucial to consider material within the context of the product concept. Picking a different material can have counterintuitive consequences. Pressurised tanks for LPG now often consist of glass-fibre-reinforced plastic instead of steel. During the certification procedure of the first of these products the assumption was that plastic could burn, so the tank would be unsafe. Instead, the final fire test showed that the plastic melted before an explosion could occur and the gas simply burned: it was much safer than steel.

Steel, as a metal, belongs to a different family than glass and plastic. Material families are

LIMITED AVAILABILITY – FUTURE RISK TO SUPPLY

RISING THREAT FROM INCREASED USE

SERIOUS THREAT IN THE NEXT 100 YEARS

The periodic table of 'endangered elements' (source Mike Pitts / the Chemistry Innovation Knowledge Transfer Network)

described further on. Materials can be combined in various ways, according to recipes and design: be aware that the right material is sometimes not just a matter of choice.

C. MATERIAL DESIGN

Existing materials may need adaptation to adjust properties. That is one reason why the sheer number of different materials is huge and continuously on the increase. Moreover, it can be necessary to develop a new technology, a new processing machine, to produce a material that can do precisely what you want it to.

For instance, the Italian company Innventia developed stretchable paper for packaging, or rather a machine that produces it. This could be

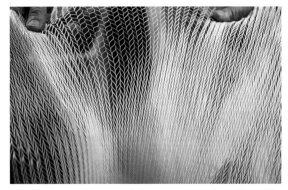

Across the world research teams are experimenting with the possibilities of translating paper-cutting techniques into new kinds of transistors and batteries that can bend, twist and deform without affecting overall performance, which may even increase battery life.

the first snippet of an idea for fast fashion. In any future, this principle demonstrates that a material cannot be seen separately from the way in which it is made.

When it comes to processing in the context of flow, the material concept must include the opposite of making, to ensure it can continue to be applied again at the highest possible value. Blending is a disadvantage, and therefore needs to be weighed against benefits for sustainability. Technology concerns decomposition as well as production. As indicated earlier – in the technical sense, recycling is always possible - 'unmaking' is doable, but the more important question concerns feasibility on a relevant scale. The best example here is laminates of plastic and metal, currently often used for food packaging. Such a material structure contributes to preservation in a flexible pack, which saves space during transportation and storage, but is very difficult to control as household waste, and to decompose for recycling on the scale in which it is produced. It must have cost considerable effort to develop the lamination process. Maybe the next step, which could be to develop a process in which the layers are not laminated anymore, but applied to packaging separately to enhance recyclability, will be just as difficult: a new process would require soup, or crisps, or whatever treat, to be packed in several consecutive steps. Or maybe these snacks would come in large containers, from which consumers could tap what they need.

MATERIAL FAMILIES

Even an old-fashioned encyclopaedia would not be sufficient to name and describe all existing materials, since new ones are born every day. It would be a faulty approach anyway, all one needs to understand is that there is a limited number of material families, each of which define a group of properties. The following overview includes some groups, or subfamilies, that are not necessarily suitable for products that flow. They may, however, contribute to the structure of products that last, which rather secretively also flow. Moreover, business models and design strategies can imply slowing down flowing products to the extent that they become lasting ones. So, meet the families:

I. Metals.

Metals are the first materials that had to be produced, by extracting them from rock. They all start out as elements from the periodic table and can be mixed, to a certain extent, with other metals and non-metallic elements into alloys to change their properties. In practice, all metals are alloys.

The most common metal materials are based on iron and aluminium, which both come in numerous variations. Steel contains carbon and varying fractions of vanadium, chromium, nickel and other elements. It can be cast, forged, rolled into sheets, drawn into wire, cut in several ways and built up through printing. Welding is steel is common practice. A disadvantage of most steel alloys is that they easily corrode. Coatings of various kinds can prevent this. Metallic coatings work very well, such as tin for cans and zinc for car plating. Steel cans currently are mostly covered in a very thin layer of polymer resin.

Aluminium can undergo the same treatments as steel. Per unit of volume it weighs about one third of steel, and its stiffness is about one third too. If it is pre-stressed with tension, like you have in a can filled with some carbonised drink (the whole concept), this doesn't matter. An advantage of aluminium is that it oxidises immediately and the resulting very thin layer of 'rust' protects the pure metal underneath. In recycling this turns into a disadvantage, particularly in foil, since alumina fractions contaminate the aluminium flow.

Separating steel and aluminium from waste is relatively simple, because of magnetism. Steel is magnetic by itself, but aluminium is a good electricity conductor and will turn magnetic through eddy currents that emerge from induction in an alternating electric field. Recycling metals into fresh raw materials is everyday practice.

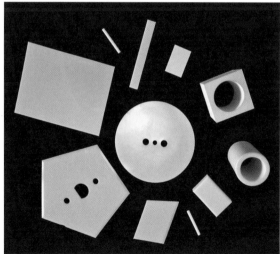

II. Ceramics.

Ceramics are second on the list, because they are metal compounds, most with oxygen, but sometimes with other elements as well. There are two basic kinds of ceramic materials, to be distinguished by their physical definition. Firstly, ceramics can consist of a stable crystalline structure. Ceramics are very strong under compression, but they also tend to be brittle. Granite and marble are good examples. One mines these. Earthenware and porcelain are far more versatile. They can be cast or formed, by

The most common ceramic material applied in packaging; amorphous, strong and recyclable.

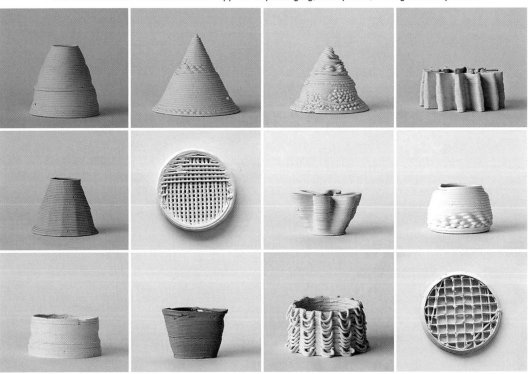

3D printed ceramic experiments by Dutch designer Olivier van Herpt. By introducing elements of randomness he intends to reintroduce error, a human touch, stochasticity. The process craved some serendipity and joy through intentional failure. Searching for repeatability and precision he found he also needed mistakes.

hand and by machines. Printing in 3D is also an option. In principle this is virtually the same as what prehistoric man did with clay, except that it is more precise. After the shaping process objects need to be fired in a kiln. Generally, few crystalline ceramics are applied in products that flow, although obviously they can be found in gadgets and gifts. Ceramic waste does exist, certainly if we include concrete.

Recycling baked ceramic material is tricky. This is not because it is difficult to do: the material can be chopped into small bits that can be added to fresh clay or porcelain as filling material. The question, however, is if using broken material as a mere addition can be seen as recycling. There is also an option to mix a ceramic (granite for example) granulate with a polymer. There we enter the area of composites, further on.

The second physical definition is defined with the word amorphous. It means that the solid mass is not really solid, but rather fluid, in an extremely viscous way. The speed with which it flows is measured in millimetres per century rather than per second. It implies unpredictability in behaviour, you never know precisely when and how it will break. A scratch may help. This is of course glass. It is very strong, stronger than steel and used in vast quantities for bottles. In 2014 it was in fact the second most used packaging material by weight.

Production is a matter of blowing a preformed hollow lump of molten glass inside a mould. A great advantage is that glass is easy to recycle at high quality and with only about 15 percent of the energy needed for fresh glass. In Europe this happens at a steady 74 percent rate. Countries such as Belgium, Slovenia or Sweden, with excellent separate collection systems, outperform Europe's average to over 95 percent. Outside Europe it is still a long way to loops so well closed.

A disadvantage of glass is that you need much of it to function well. In that sense it matches the earlier comparison of concrete with aluminium. Therefore, transportation costs are high. PET has a significant edge over glass in this respect: extremely thin-walled bottles (in the range of 0.5 millimetres) weigh next to nothing.

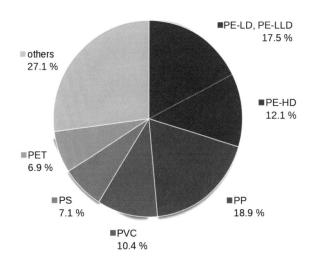

Commodity plastics are plastics used in high volume; they exhibit relatively low mechanical properties and are of low cost.

III. Plastics.

With plastics we arrive at the family of materials with a considerable contribution to make. So many are travelling around and being processed and ending up in parks, forests, landfills and the ocean, that they are probably the most important messenger to tell us that things have to change. There is, however, more to plastic than meets the popular eye. It is not simply a category with a bad press. There are many different family members and, believe it or not, many of them have advantages. They all consist of large carbon-based molecules, which renders them 'organic', totally counterintuitively from the word being almost synonymous with 'natural'.

1. Thermosets is the name of the first group. They consist of interconnected long molecules and are characterized by the fact that they originate from a chemical reaction that starts by bringing together reacting fluids, or by putting the unset material under the influence of light, or pressure and heat. Thermosets are quite strong and stiff, and they can withstand different influences. Melting is not what they can do. Heat can distort their integrity and they may burn, but they won't get softer and turn liquid. This makes them comparable to crystalline ceramics when it comes to recycling. Shredded thermosets can be applied as a filler. There are chemical ways to divide them up into different compounds that can be reapplied. A large-scale return to the original thermoset resin, however, is as yet close to impossible.

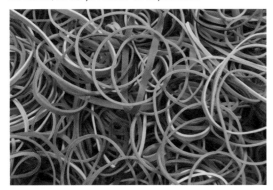

2. Rubbers are very soft and elastic, when compared to thermosets. They are produced by chemical treatment of latex from rubber trees and quality is enhanced by adding sulphur under high pressure: vulcanisation.

Since car tyres are made of rubber, the material is omnipresent. Car tyres have evolved into sophisticated compositions of steel wire, layers of fabric and rubber. There is a system to 'reemploy' the tyre materials after use, including even the minerals they picked up from roads. These tyre materials end up in different applications, such as roof tiles that can hold water for plants. Experiments are going on to 'de-vulcanise' rubber, which would imply that sooner or later rubber compounds could contain 50 percent recycled rubber. Incidentally, the word is derived from the verb 'to rub'. It was first used to wipe out mistakes made with a pencil. Synthetic rubbers exist as well. In fact, the 'rubbery phase' is a physical condition that all thermoplastics possess within a certain temperature range.

3. Thermoplastic's long carbon-based molecules are not fixed in complicated net structures, like thermosets and rubbers, but they are loosely arranged along and around each other so they can 'move' a bit. They have an amorphous character, some kinds more than others, which makes them behave somewhat like glass. They also get more transformable when they are moderately heated, certainly when compared to glass or steel. They don't have a precise melting temperature, like steel, but a melting and setting trajectory.

There are many ways to 'freeze' them in a certain shape. They all start as grains ('granulate') that are melted. Next the hot plastic is extruded to become profile, or pipe, or plate. Pipe can be inflated to become foil. Plate can be used to be heated again and formed in moulds with some air pressure: thermoforming. Many packaging trays are made in this way. The procedures to add all kinds of pictures and texts to the surface are manifold and quite sophisticated. The most precise shaping technology is injection moulding: high mechanical pressure injects melted plastic in a mould that defines the product. Some moulds are very advanced (and expensive). Packaging for prestigious perfume brands may consist of different colours of material that in the mould are transformed into very complicated structures, which for instance include a relief of text of one colour against a background of a different hue: all to prevent the packaging from being copied.

There are different kinds of thermoplastics, thousands if you just look at formulas and

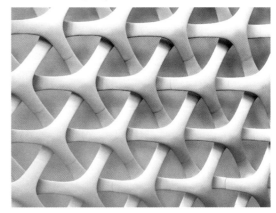

recipes. Here we limit ourselves to three main subfamilies:

POLYOLEFINS is the collective name for two soft and versatile, tough polyethylenes and polypropylenes. They are used in gargantuan amounts, particularly for easy packaging. Think plastic bag and you understand that they are all over the place. The subdivision is not entirely strict. There is an extremely pure type of polyethylene, with well-arranged parallel molecules, that is known as Dyneema, one of the strongest fibres on earth.

Recycling is possible, as in all cases dependent on flow purity, which is hard to achieve. Collecting, or rather controlling them to a sufficient level, is difficult.

STRUCTURAL THERMOPLASTICS, the most important of which are PS, ABS, PVC, PC, PMMA and high-grade types, like PEI, PEEK and PPS, are robust and used for an enormous range of different functions, packaging and small products, like coat hangers and office accessories. However, they are also used to define the body of electric and electronic products, car parts and airplane parts. The mass-produced kinds contribute to waste as well and are difficult to distinguish from each other.

Gap exploiting designers often turn them into new products by adapting them to the aesthetic conventions of design products. They like to use PET bottles, but these happen to be quite well recyclable, due to the existing retrieval systems. There is no harm in plastic waste design, often meant as a statement in favour of recycling and continuse, but its contribution to plastic waste reduction is symbolic.

BIOPLASTICS differ from the other thermoplastics in that they are not made from crude oil. Their carbon base is derived from plant materials. Again, there are different kinds with different grades of degradability, which is their alternative to recycling. Some are made from carbohydrates and for instance used as trays for snacks. These can dissolve in water. PLA (Polylactid acid) is currently is the most important one. It is compostable, given the right industrial conditions. Bioplastics over the years have had a hard time proving their right to exist. When they appeared, the general feeling was that their ability to dissolve should be functional. In slaughterhouses for instances they use dissolvable plastic stoppers to keep all the moisture inside carcasses temporarily. In greenhouses they use them for clips to tie up tomato plants. Bioplastics used to be mistrusted because of their flow contamination risk. This is gradually changing. It could be that their ability to disappear is turning into a functional asset to reduce the amount of energy needed for recycling.

IV. ORGANIC MATERIALS

Organic materials are the ones that grow, and are used as they are, except that they may be cut to size or stuck together. Like their bioplastic cousins they can contribute to soil fertilisation. An important trait of character is that they possess a structure that defines their structural potential.

- *Wood* for instance has originally grown as a part of a tree, which has defined the size, direction and layering of the cells it consists of. Plants often have a fibre structure that can provide strength. Flax, hemp and, particularly, bamboo are well known in this respect. Plant fibres can be strong, but not as strong as continuous glass or carbon fibres. The reason is that plant and tree fibres consist of segments, cells, that have grown attached to each other. Trees take many years to grow, which contradicts applying them for products that exist briefly. On the other hand, paper (next) comes from trees and can be recycled many times. Because wood and plants have been in use for ages, there is a lot of experience with combining them and turning them into houses, chariots and different functional objects.

- *Paper and cardboard* are composed of wood pulp, which is both recyclable and compostable. Production requires harvesting trees and drying the pulp takes energy. Bleaching may unleash chemicals. Because of their origin, these materials are seen as a cause of deforestation. This is also the case with cellulose fibres, which have received growing attention in fashion

(Lyocell). There is a shift going on. Agricultural materials and fast-growing bamboo are replacing wood, to produce cellulose fibres.

On the other hand, paper and cardboard can be recycled quite well, which happens on a considerable scale, although this is less simple than one may think: there are about thirty different kinds and they have to be selected and kept apart. Of course, contaminated pieces have to be removed. Recycling one ton of paper saves around 16.3 barrels of oil, 26,500 litres of water and 17 trees. Paper cannot be endlessly recycled: around seven times at the most, often with toilet paper as the final outcome. At the end, the fibres are too short and no longer have sufficient strength. There is an option to mix different fibre qualities.

Soaring e-commerce is causing an increase in the use of cardboard boxes, so much so that in some places it disturbs waste flows: cardboard ends up with normal waste, missing out on recycling potential. More control over the paper and cardboard flow is turning increasingly urgent.

V. COMPOSED MATERIALS

Composed materials are structures that consist of one or more materials and in common conversation are just named materials, like wood. In many instances these structures are bespoke and have been designed for a specific purpose. Being built up as a structure in some subfamilies complicates their 'unmaking' and recycling, while in others it is no big deal.

- *Laminates* are the least complicated kind: thin layers of different materials, or of similar materials with slightly different compositions: virgin on the outside, regenerated on the inside. Depending on their precise composition they can be very difficult to recycle, but the capacity of some to preserve food has to be weighed against this. An interesting dilemma: do you want to eat well-preserved enjoyable crackly crisps, or be able to easily recycle the bag they came in and eat stale snacks?

• **Textiles**, particularly in the fashion industry, are currently the most damaging trade, after fossil fuels. There are four reasons. The first is that production is overdoing it on a massive scale. By and large 30 percent of large brand clothing items remain unsold. Apparel use on the other hand is 'underdoing it'. People buy clothes and don't wear them. The third reason is that continuse is hard to come to grips with due to the volatility of fashion. It may be fashionable now (look at second hand clothes as a style opportunity) but will this observation turn into a real ongoing trend? That will take a lot of communication effort and considerable changes in business models. And then the final reason, concerning textile itself: taking clothes apart is just about as complicated as putting them together. Knitwear may be slightly easier than woven fabrics. The next option is shredding, which is likely to harm fibres, thereby reducing the number of times they can be reprocessed.

An option could be to start thinking in 'easier' materials for fast fashion, such as non-wovens and papers. Plastic fibres are also easier to recycle than cotton, and some renewable fibres, like flax and hemp, as mentioned elsewhere, are compostable. The best way to look at textiles is that there are many opportunities for flow improvement. All they need is some rethinking.

• **Composites** consist of material fragments, or fibres combined with a polymer. Fibres can be segmented or continuous. And they can be renewable, or plastic, or metal, or glass and carbon. The polymer 'matrix' in which the fibres

are embedded can be thermoset, thermoplastic and bioplastic. The most important advantage of composites is that they distribute forces in structures in such a way, that the amount of material can be reduced: strong, stiff and light. They are being applied in supercars and aircraft, but also freight containers, and gradually find their way. Recycling, for instance carbon or glass fibre reinforced composites, is gradually getting there. The scale on which they can and will be used for products that flow is unclear.

• **Foams** need to be mentioned, for they are composites of materials with air. These are used for insulation and packaging. Foams protect. There are many different combinations and many levels of refinement. In micro-foams the air mainly serves to reduce the amount of material. The other extreme consists of chains of polyethylene air bubbles used to protect products inside boxes. Styrofoam or EPS is used quite a lot for packaging, but paper alternatives are taking over. Metals can also be turned into foam, for very specific purposes. Even paper foam is an option now: harmless foam. In general, it can help to consider air as a material. When it concerns flow, life for materials is not easy, because the end of it is always in sight. No matter what kind of material from whatever family it concerns, flow implies: take it easy and slow down. Transportation and transformation shouldn't be much of a deal in a thrifty culture.

An online tool used to assess the environmental performance of material, packaging, and manufacturing facilities, as well as to assess the social and labour performance of material, packaging, and manufacturing facilities.

C&A's T-shirts were certified GOLD – an achievement level not seen before for a fashion garment.

DEALING WITH MATERIALS

HIGG INDEX: For the apparel industry, Higg provides you with a user friendly and transparent tool to compare the environmental impact of different materials. It also offers a free tool for fashion designers to design for sustainability. An example of a comparison of cotton with nylon and polyester shows that cotton has the largest environmental impact. Included are contribution to global warming, eutrophication, water scarcity, abiotic resource depletion, use of fossil fuels and chemicals. The processing of materials into garments is included too.

RESPONSIBLE T-SHIRTS: The fashion retailer C&A proved that it is possible to produce sustainable clothing at an affordable price. They partnered with two Indian suppliers to develop and produce Gold level Cradle to Cradle (C2C) Certified™ T-shirts; and brough 1.3 million pieces to market in 2017. The Cradle to Cradle Certified™ Product Standard is an independent certification scheme including criteria on raw material and usage of chemicals, designing products with materials that allow reutilisation, releasing only clean water, using only renewable energy sources, and providing safe and dignified working conditions. C&A received the "Gold" level, making it the first piece of fashion produced in Asia on a large scale while at the same time meeting the strict and extensive C2C standards.

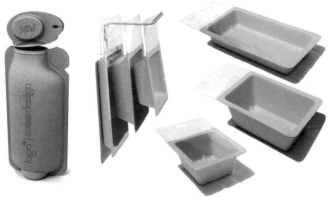

WICKED PROBLEMS: A Dutch consultancy, Partners for Innovation in Amsterdam, researched recycling problem items for the most common packaging items. Sometimes products are hard to recycle for lack of suitable separation technology, sometimes economy is the bottleneck. Multilayer laminates are still very difficult to recycle.

AIR PAPER: Mix air with paper and you get PaperFoam®. A special mixture of industrial starch, cellulose fibres and water is injected into a mould to shape well-defined trays in various sizes. Consequently, the mould is heated to bake and dry the material, and turn it into foam. PaperFoam® weighs approximately 180 grams per litre. This translates into a possible weight reduction of 40 percent compared to traditional packaging products.

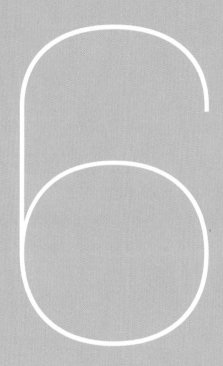

different responsibilities

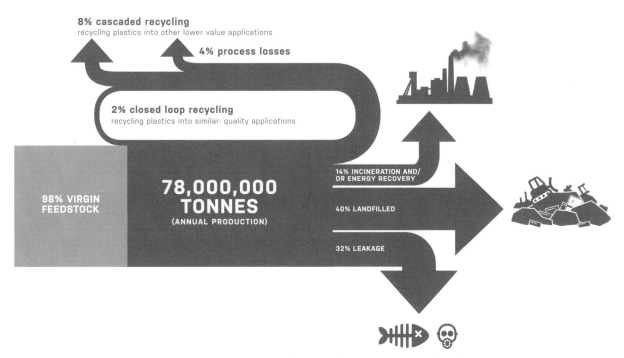

The flow of plastics in a linear system (Source: Ellen MacArthur Foundation)

Now let us start with us. We know convenience is our main driving force. That is why we eat ready meals and own things we forget to use, and why we drop bottles all over the place. Self-interest makes us spoil ourselves with long showers and loud music. Naturally, we are instinctive animals too, consuming too much and shopping for the sake of consumption. As reasonable citizens, on the other hand, we say we try our best to live responsible lives. We like to think of ourselves as sensible beings.

COMMON GROUND

Group us together in societies and organisations, governments, traffic jams, restaurants, tourist venues, stadiums, swiping screens, losing track of reality: needless to say, things become even more serpentine. Yet, in defiance of the complexity of human life, with all its collisions of interest, we have to learn how to control the ruthless flow of making, using and ditching stuff. Fortunately, believe it or not, there are signs we are beginning to understand how to deal with this abundance.

People from different professions and with different backgrounds, each within their own affinity, are beginning to shape a map of interventions, maybe even without being aware of it. Scientists observe what is happening to our living environment and scrutinise the effects of attempts to gain control. No matter how loudly they warn us, the effects of human measures to gain control over our self-created economic machine have been modest at the most. Use of energy and materials is still on the increase because affluence is on the rise everywhere. However, the naivety of the assumptions behind continuous economic growth has become apparent. Ideas such as Doughnut economics from Kate Raworth, which favour a more realistic sustainment of a balance between societal wellbeing, sufficient production and consumption,

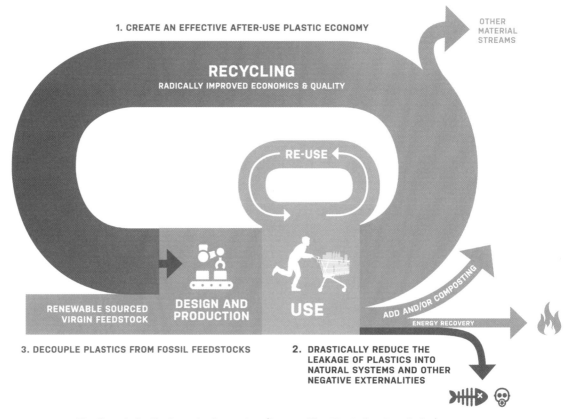

1. CREATE AN EFFECTIVE AFTER-USE PLASTIC ECONOMY

RECYCLING
RADICALLY IMPROVED ECONOMICS & QUALITY

OTHER MATERIAL STREAMS

RE-USE

RENEWABLE SOURCED VIRGIN FEEDSTOCK

DESIGN AND PRODUCTION

USE

ADD AND/OR COMPOSTING

ENERGY RECOVERY

3. DECOUPLE PLASTICS FROM FOSSIL FEEDSTOCKS

2. DRASTICALLY REDUCE THE LEAKAGE OF PLASTICS INTO NATURAL SYSTEMS AND OTHER NEGATIVE EXTERNALITIES

The flow of plastics in a circular system (Source: Ellen MacArthur Foundation)

herald new potential for change. Capitals other than just the material are being discussed and defined.

Engineers tend to consider the side effects of material and energetic 'overflow' – let us call it that – mainly a technical problem that can be solved by innovation of products, transportation and recycling procedures. With companies and designers, they focus on the concept of circularity, in which ideally all goes around in a perfect circle of material utilisation. Here the natural ecological circle, in which ideally all interventions together allegedly produce a zero-sum output, is the ideal metaphor of this way of attacking wastefulness. This very book is created along these circular lines, but not without awareness of the fact that the idea of circular economy as a benchmark is merely a stage in methodological development. Circular economy does not sufficiently address the inevitable imperfections of resource exploitation and waste production, nor the continuing increase in energy consumption. Nevertheless, the circle is a strong metaphor to bring people together to start

working on a tremendous amount of ideas to adjust the systems we have developed.

MOVING AROUND IN CIRCLES
The largest unit of human organisation is government. It is not easy to cover all about governments in one sentence but let us have a go. Governments function on different levels of scale: local, regional, national and supra-national (European). Depending on place and time, they have different kinds of authority, which depends on a continuously shifting balance between population preferences, capital, individual power and charisma, job security, lobbyism, legislation, violence, paranoia, idealism and political skills and certainly more.

Governments behave opportunistically. They act on support through political outcomes, which are difficult to predict. Timing is crucial but coincidental. This implies that there may be a decision, for instance to prescribe mixing a certain percentage of biofuel with fossil fuel, which at first seemed beneficial to emission reduction, but after a few months proved otherwise. In addition, it would go at the cost of soil for food production. Soon the measure had to be abolished. This all happened in Europe in 2007. Government decisions, however, depending on the political situation, usually demonstrate good intentions, for instance by forbidding plastic bags and drinking straws. They do tend to concern symptoms instead of causes. After all, the best way to start emitting less CO_2 is to diminish energy dependence, a challenge of innovation that governments tend to avoid, because it is thought that reduction hampers growth.

All the aforementioned plus everyone else, no matter what they do, participate in society. That is where public opinion, beliefs and behaviours emerge from events, alternative facts, news, experiences, facts and preferences. Society is chock-full of them and there is a general idea that somewhere over the rainbow lives an entity, which is The Solution. Supposedly, it is easy. All we have to do is take a spoonful of creativity and find it.

If only it was as simple as that. There are always reasons to hesitate about change for the better. One kind could be named 'current knowledge'. It looks like this: "If we would partly replace fossil plastic packaging with bioplastic packaging, the recycling flow world be contaminated. Therefore, bioplastics are a bad idea." This way of reasoning hampers long-term development, because it neglects the potential to find new answers now and try and optimise them for functioning later on.

Probably the most important inhibitory effect on diminishing the environmental effects of the flows we produce is the question who is responsible, which virtually always ends up in waiting for 'the others'.

At the most, an individual will feel somewhat guilty about drinking soda pop through a plastic straw, but the café manager is really held responsible. As a consumer, one feels powerless, because of the idea that, for instance, to continuse one second-hand blouse instead of buying something new, won't change anything.

Already there's a dilemma too. If you save money by purchasing a used item, you will have to decide on an alternative to spend it on. It could be a responsible fashion innovation, so you must wait until some company launches that, or until

In 1941 the British gouvernement lanched a campaign to encourage people to take good care of their clothes and mend them when they became worn, rather than throw them away and buy new ones.

government rations clothing. As a matter of fact, this happened in the 1940s in the UK, when scarcity ruled because of the war. Government took control over the British fashion industry by issuing *Utility Clothing*. This is a brilliant example of how shared interest can create support for vast adaptations to circumstances: in this case scarcity through war. It may well be that currently we should not be looking for The Solution, but for an urgency that is close and concrete. Searching is wiser than waiting.

Trade and industry, in their turn, tend to wait for markets to make responsible choices. They offer responsibility and greenness as a feature and react on numbers of people that use their products. For most customers, on the other hand, behave more on intuition than on awareness. Greenness to them is important as far as it is within what could be named 'the atmosphere of normality'. They buy green 'because everyone does'. This choice does not imply sensibility.

An interesting example is gluten. About one percent of people develops coeliac disease and about five percent is sensitive to gluten. Yet, the market for gluten-free cereals currently amounts to about 30 percent, because people identify with healthy food. Producers gladly comply. Apple's Steve Jobs said: 'It is not the consumer's job to know what they want'. This is the reason companies are the strongest implementers. Nevertheless, they come with proposals and wait for developments in 'the

The perfect gift box for people who already have everything (Hurray! You got nothing) available for € 25 or € 35. The money goes to Greenpeace for ocean and forest protection and bans on pesticides.

There is a trend towards organic, healthy and high-quality food. in 2016, the organic food market reached $43 billion and currently represents 5.3 percent of total retail food sales in the U.S.

atmosphere of normality', which emerges from an image produced by news and social media perception and gossip.

Government measures are a strong impulse for company policy change too and, in their turn, they come forward from the described mix of knowledge, idealism and political opportunism. Government interventions, as said earlier, usually address symptoms. They depend on support that may be the consequence of whimsical public preferences influenced by media coverage and strong images. A seahorse holding a cotton stick can amplify attention for plastic soup a thousand times and enforce government measures to prohibit plastic drinking straws. This happened to be just the thing that traders and consumers were waiting for to make a good impression and it helped pull behaviour to be careful with plastic waste into normality, just like smoking is being pushed out.

This is one of many positive signs. Whereas 'the environment' used to be considered a leftist and sometimes even anti-humanist theme, it has now become a world-wide concern. Overlooking the way things are gradually changing reveals that there is common ground, on which there is room to design and introduce less damaging flows and to develop and implement lighter and more precise systems that no longer exhaust resources. This is not a matter of The System that we can change once we have The Solution.

If human society is a system, it is chaotic, in the mathematical sense, which means that its development is unpredictable, with influences from all directions coming in, but that there are episodes of relative stability. Therefore, society is certainly not insensitive to initiatives that together have the potential to change the stable regime of normality towards a less hazardous future. Proactivity on this common ground is crucial.

CONSUMER INFLUENCE

Not everybody is prepared to wait for others, even in citizen circles, which are allegedly powerless. Individuals through social media put pressure on governments to act. And they did.

Some have embraced a life without packaging and prove that it is possible to expel packaging waste from their lives. Emily Lowe, from The Netherlands, wrote a book about Life without Waste, in which she explains how she and her family succeed in producing only one small bin of garbage a month. She also keeps a blog on waste reduction. Tactics range from continue using your own disposables to making your personal healthcare products.

Another example is zero waste advocate Bea Johnson in the USA. She wrote 'Zero Waste Home: The Ultimate Guide to Simplifying Your Life' and created a 'bulk finder' on her website· It helps consumers to find shops that sell goods without packaging. These kinds of opinion leaders have an increasing popularity among committed consumers.

Online food delivery services are growing worldwide.

Governments all over the world have responded and taken action to ban sales of plastic bags or charge for them in various ways. The Bangladesh government was the first to do so in 2002, imposing a total ban, followed by other countries such as Rwanda, China, Taiwan and Macedonia. Some countries in Western Europe impose a fee per bag. As a consequence, the market for reusable shopping bags is growing.

Plastic bags mark a beginning. Many companies started to experiment with accepting reusable cups with discounts. All over the world, civil society groups and citizens press for bottle deposit schemes and a clean and slave-labour-free fashion industry. Even party balloons have become doubtful because left to the vagaries of the wind they end up littering the environment.

CONSUMER BEHAVIOUR

Changes in consumer behaviour are clearest in what people eat and where they get it from. There is a definite shift towards organic food. This marks a rising environmental awareness linked to health anxiety. It has to be said that 'organic' and 'sustainable' are not synonymous. Particularly with fish the labels are confusing. Wild fish by the rules of the USDA (United States Department of Agriculture) may not be called 'organic' and farmed

fish can be 'organic', because of the way in which it is farmed, which is not necessarily sustainable. Salmon are fed with other fish and coloured pink through natural pigments from shrimp or yeast, or through synthetic astaxanthin. Anyway, apart from glitches in procedures and definitions, the market for well-considered food in the US doubled in ten years. In Europe developments were similar.

Food delivery is growing tremendously. Worldwide the market is now around 100 billion dollars and expected to keep on growing.

It is quite interesting that businesses use disposable packaging and that 80 percent of customers hardly ever change platform. This implies that there may be a valid opportunity to start using non-disposables. When the next order is delivered, the client hands back the containers of the previous one.

LUNCHBOXES THAT LAST

In the Far East, particularly in India, it is quite common for women to prepare a warm lunch for their husbands at work. The men take the metal lunchbox, or tiffin, or dabba, with them, but there are specialised delivery services too. In Mumbai, in

This man contributes to a service that delivers hot lunches from homes, restaurants and central kitchens to people at work.

the North West of India, they have a now legendary system. Delivery persons, called Dabbawalla, pick up hot lunch boxes from the homes of office workers, deliver them to the right place, and return the empty boxes afterwards. Each and every weekday about 5,000 people deliver up to 200,000 lunches travelling by train and bicycles, faultlessly and on time. Statisticians claim that this is impossible, but it happens anyway. It is not clear if the Dabbawalla will keep on performing with continuing economic growth. Their wages are low and translating the service to different regions with different customs is not self-evident. Emancipation may also endanger the service. Still, it sets a good example of combining services with lasting equipment. There are attempts to make services more universal.

CHANGING MARKETS

Big accidents have happened in fashion. Two brands made the news because they felt forced to destroy considerable amounts of money worth of unsold clothing. This may be a tip of the iceberg, but no matter what: it is a symptom of a changing

market. At the end of the 1990s the engine of fashion changed from branded seasonal collections of well-known designers that proposed silhouettes, colours, patterns and details, to a fierce competition of brands storming the market with brands offering items that were directly inspired by incidents, and images of the rich and famous at an increasing frequency.

Currently high-speed fashion seems to have reached its peak. Awareness of the absurdity of speed is growing, together with the recognition of the value of second-hand clothing.

China is mainly producing at high speed for the home market now, and second-hand items are exported to African countries, where customers are prepared to pay more for the used clothes than for new items, because those have better quality. In Europe and the US, trade on resale markets is soaring. It could be the announcement of an enduring change. You're never sure with fashion, but this is an emerging expression of preference for contin-use over fast fashion.

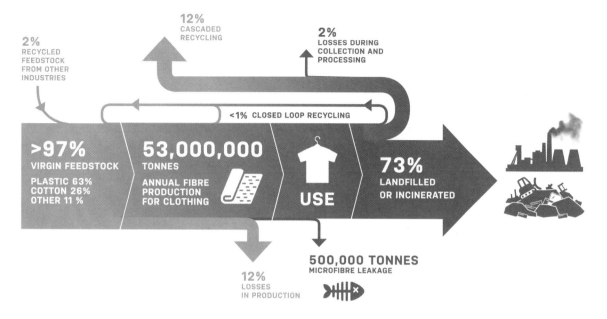

The flow of textile in the linear system (Source: Ellen MacArthur Foundation)

Similar developments can be observed on other markets. In an interview in the Guardian, Steve Howard, IKEA's head of sustainability talked about 'peak curtains' as the moment in time when a sufficient amount of furniture has been produced. He said the enterprise is increasingly trying to help customers to live in a more environmentally friendly way.

GOVERNMENT ATTRACTIONS

Governments give direction to adaptation. They do this by providing subsidies to activities they favour and tax objects they qualify as undesirable. Precise rulings or laws can prohibit or prescribe procedures altogether. Government measures can provide the momentum to set change in motion, from neighbourhoods and municipalities that stimulate concrete activities, via national governments that bring out policies and rulings and express abstract multilateral intentions at global conventions, such as the UN. They defined 17 goals for sustainability, of which *responsible production and consumption* is just one. Government measures reach from a village gardening competition to national representatives travelling to some well-known city to sign a shared policy promise.

Over the past years, many measures have been taken to move society into the direction of sustainability. Wind and solar power have made a breakthrough and electric vehicles are on the verge

of becoming normal, whereas only a few people know that the first car to break the 100 km/h limit was the electric rocket shaped *Jamais Content* (Never Satisfied) in 1899.

There is a Chinese saying, wrongly attributed to Benjamin Franklin, that says: tell me and I'll forget, teach me and I'll remember, engage me and I'll learn.

The local *level* is very much suitable for the latter. Children learn from engaging in community activities that make sense, such as collecting plastic litter, which happens all over the world and raises awareness of us spoiling our surroundings. It also shows something that simply shouldn't be necessary if flows were mastered.

Apart from deposit schemes that have been mentioned earlier, there are also experiments going on with procedures to help citizens separate their waste. WASTED is the name of a project supported by the municipality of Amsterdam. People can separate their waste and 'post' it in the appropriate container where they can scan its QR label and take a picture. Participants get rewarded for their effort with discounts in local shops. This calls for a critical question: is a discount a good incentive? In this case it implies waste separation at the cost of consumption, which may produce more waste.

FOOD WASTE	ENVIRONMENTAL WASTE	FINANCIAL WASTE	CLOSING THE FOOD GAP
			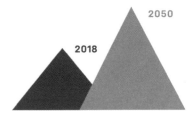
24%	**198,000,000**	**$1600**	**69%**
Calories produced for people that are never consumed	Hectares used to produce food we don't eat (about the size of Mexico)	Value of food thrown out by the avarage U.S. family per year	Required increase in food calories needed to feed 9.6 bllion people by 2050

Global food loss and waste: By 2050 the world will need about 60% more calories per year in order to feed a projected 9 billion people. Cutting the rate of global food loss and waste could help close this food gap while creating environmental and economic benefits (source World Resources Institute).

Albeit on a smaller scale, this is very similar to winning an airplane ticket in an energy use reduction contest, which happens to be common practice in some energy companies. It exemplifies the notorious 'rebound effect' that is always lurking around the corner spoiling good ideas: helping the environment save money that afterwards can be used to do the opposite. The rebound effect needs to be tamed.

Maybe it could be taxed. On the *national* level, taxing pollution is done hesitantly, but is not uncommon. It can force companies to gain control over the waste they produce. There are fields where tax or tax discounts on repair and maintenance labour can contribute. Sweden is experimenting with a VAT reduction on repair. In terms of flow, maybe companies that specialise in flow control could be allowed to pay less. The point as always is flow containment.

Other measures are to support supply chain projects where packaging producers collaborate with recyclers and discuss how to improve the recyclability of their products. Different initiatives use online tools to match the supply and demand of excess or recycled materials, or to help companies with creating products with value that can be cultivated in time.

Food relates directly to public health and is therefore more susceptible to regulation concerning sustainability. Soil, water and biodiversity, but also minerals such as as nitrogen, phosphate and trace elements, must be effectively managed and used. Reducing food waste can be achieved by enhancing company transparency concerning residual flow and changing consumer behaviour. Throughout the food chain, consumers by far contribute most to waste. France is no longer allowing supermarkets to throw away food. Germany is focusing on reforming the expiration date definition.

Interestingly, a diet with less processed food reduces waste production: health and waste management go hand in hand.

Industrial food processing creates residual flows, both during production and after consumption, partly consisting of packaging material, and requires a higher processing energy input. Some countries have policies to put higher VAT tariffs on processed food. France does that to sweets.

A reduction of animal protein in our diets is a very effective way to reduce waste. The government of China is now campaining to reduce the amount of meat eaten by its citizens.

In general we need to decrease environmental pressure on ecosystems and try to exploit residual streams, such as tomato stalks, beet pulp and stale bread rather than just dump them, to make sure biomass is not lost.

Closing loops really gets within reach through schemes that promote *Extended Producer*

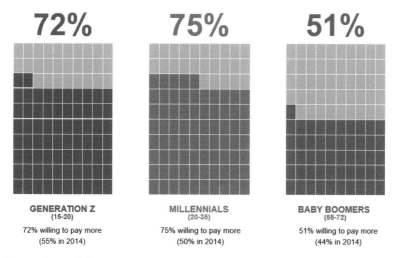

72%
GENERATION Z
(15-20)
72% willing to pay more
(55% in 2014)

75%
MILLENNIALS
(20-35)
75% willing to pay more
(50% in 2014)

51%
BABY BOOMERS
(55-72)
51% willing to pay more
(44% in 2014)

Sustainability is not only a call for environmental protection, but also a customer demand. 58% of consumers report taking a company's impact on the environment into consideration when purchasing, and almost three-out-of-four Millennials (73%) are willing to pay extra for sustainable offerings.

Responsibility. These imply that producers are responsible for what they produce all the way to the end. In the Netherlands, producers and importers of packaging pay a fee for waste collection and innovations around recycling. They pay per kilogram. This scheme is organised by the 'Packaging Waste Fund', which was established by producers and importers to meet their responsibilities. This is a collective version of the scheme.

Industry collaboration can be governmentally imposed or supported. The European funding programme Horizon2020, for example, requires industry collaboration if they apply for funding projects. Moreover, the European Standardisation Committee is developing quality standards for sorted plastics waste and recycled plastics in cooperation with the European Standardisation Committee: making it easier to apply recycled material in new products.

Extended Producer Responsibility may also apply to textiles, which happens in France. Every company that introduces clothing, household linen and footwear items to the French market must either set up its own collecting and recycling programme, which must be accredited by the authorities, or pay a contribution to an organisation called Eco TLC (TLC is textile; lingerie; chaussure: textile; laundry; footwear). The amount derives from the previous year's volume put on the market and the size of the items. Moreover, all products that contain recycled fibres receive a discount that depends on the

percentage. Here we see a choice between a private and a collective version.

One step further than taxing is a ban, like the one on plastic straws. In Japan, PET bottles have to be completely transparent. When lamination or coatings are required, evaluation for recyclability is obligatory. Apparently, national governments can do this. Bear in mind, however, that recyclability can change over time due to new technological developments. Evaluation is not a one-shot procedure. Recycling technology is likely to evolve and become economically viable.

TECHNOLOGY OPPORTUNITIES
Being human we have a knack for tools. We have been making and refining them for ages, particularly to make things that fulfil our every wish at the soonest. In the recent past we have discovered that there is another side to making that we will have to learn as well. It could be named intelligent demolition: destroying in such a way that fresh materials are produced. It is the area where the most promising developments are on the verge of breakthrough.

We start paradoxically with 3D printing, because of its projected recycling potential. It is one principle of what is known as additive manufacturing: building objects by adding small bits of material. From the mid 1990s onwards, various versions appeared on the market. Because of its miraculous way of

Danit Peleg is a Fashion Designer. Her Liberty Leading the People collection made in 2015, was the first fashion collection in the world to be printed entirely at home using desktop 3D printers.

producing something out of nothing, it quickly rose to popularity. 3D printing set high expectations, which related to single material home production and easy recycling. As a matter of fact, a high-end department store chain put small 3D printers in their shops for Christmas in 2012. They may have sold some, but the machines did not return the next year.

This may be the consequence of limited convenience. A buyer can easily download some model from the Internet and print it in plastic, but this takes time and the result is not really exhilarating. For consumers a 3D printer is a complicated toy, like the home computer was in the 1980s.

3D printing is still in its infancy but developing very quickly. Many materials can now be printed in more directions than just the three Euclidean axes. Combinations are possible with milling and drilling and sanding tools. And although materials have to be printable – most of them still aren't – composing materials is possible, for instance with 'fibre-placement' that can build fibre-reinforced composite structures.

The most important feature of this technology is that it is flexible. There is no mould and shapes can be modified from one to the next by changing a few numbers in an algorithm. Automatically of course. That is the reason why 3D printing is suitable for products that tend to vary in shape and size, like shoes, from one material that can easily be recycled.

3D trainers

3D printed shoes are now available from Nike, New Balance and Under Armor, but at very high prices. At a certain point in its development a technology, such as 3D printing, in itself can be a selling point, a matter of identity rather than technological or functional benefit.

Adidas wants to scale up 3D printing. They're testing a fully automated 3D printing and robotics 'Speedfactory' in Germany and they're planning to open a second one in the US.

The mid-sole of the Futurecraft 4D shoe is created with Continuous Liquid Interface Production, in which the polymer is pulled out of a vessel of liquid polymer resin and 'frozen' into its desired shape with

3D printing is not only an insanely effective technology allowing to create footwear that isn't just unique or difficult to counterfeit, it also means designers and scientists have zero constraints. As shown by the Adidas Grit shoes.

The RFID technology can positively impact everything from sales and online cart abandonment to pricing in-store customer services and returning logistics.

ultraviolet light. This principle stems from the early principle of stereo-lithography, the first principle that became available.

Carbon, the Silicon Valley company that developed it, claims the method is faster and more adaptable than traditional additive printing. It is a promising start, but as of now UV-cured polymers are not yet easy and affordable to recycle. Tricks are being developed but they may slow down production. A better option can be to somehow speed up thermo-plastic printing. Usually this kind of dilemma gets solved. Once an innovation goal is clear, problems disappear in due course.

The current belief is that trainers come in shoeboxes. This may change in future. What certainly will change is packaging traceability and interactivity. The current terms are active packaging, intelligent packaging, and smart packaging. They're used for foods, pharmaceuticals, and several other types of products, to help extend shelf life, monitor freshness, display information on quality, improve safety and enhance convenience.

Active packaging implies active functionality. Intelligent and smart packaging involves the ability to sense or measure some product attribute, such as the inner atmosphere of the packaging, or the shipping environment outside it, or simply its size. This information can be communicated to users or trigger active packaging functions. An example of

smart packaging is a food container that shows the time expiration date, depending on what the product inside has gone through. The German Ministry of Agriculture is subsidising smart packaging with 10 million Euros, to help rule out food waste

Radio food

RFID (Radio Frequency Identity) tag systems work with chips that store data about the products to which they are attached. The advantages over inventory systems that rely on barcodes are that RFID chips have a unique identity, can store extensive information, can be read at greater distances, work quickly and don't require a direct line of sight for scanning.

Avery Dennison RFID has been exploring the techno-logy's potential to address the issue of food waste. There is a significant level of food waste in the supply chain due to inaccurate information. In stores it can easily happen that packages are offered to customers in the wrong order, thereby overruling the expiration date. RFID accuracy and readability can prevent this from happening. Apart from that, sustainability-conscious brands see potential in letting consumers know what to do with an item at the end of its life.

Smart separation

Demolition starts with getting the right stuff on the right place by sorting and logistic procedures. The

Unilever has announced a partnership with start-up company Ioniqa and the largest global producer of PET resin Indorama Ventures to pioneer a new technology which converts PET waste back into virgin grade material for use in food packaging.

first step is to reframe all the products we have made as a mine. That's how we arrive at the name Urban Mining Corp of a company in Rotterdam, the Netherlands. They build machines for a new separation technology in which the sinking and floating properties of different materials can be recognised. The process fluid consists of a water solution with ferrous oxide (rust). When this liquid is placed above a magnet, it is affected by both the magnetic field and gravity.

As a consequence, the fluid density varies from dense, where the magnet is closest and most ferrous oxide is attracted, to virtually waterish. Now material particles trying to find their way in there, will float higher when their density is lower, according to the density of the material that they consist of. So, for the sake of the explanation let us suppose that near the magnet the fluid has the same density as iron. Aluminium particles will float just above that, but certainly not as high up as polyethylene, because that will float on top, having a density similar to water: the higher above the magnet the lower the density of materials, which is a grand opportunity to separate them.

So far, the company has developed machines for packaging and technical plastics, heavy non-ferrous metals, diamond tailings and pigments respectively.

There is also a version for chemical analysis and very small batches.

Instead of mechanically separating the different plastic types – even fiddling with magnets is mechanical – it is also an option to use a chemical process to recycle mixed plastics. This will bring plastics back to their original building blocks, monomers, or even naphtha, the fragment of crude oil that can be turned into monomers. The benefit is that you actually bring back the material to its original quality. However, the technology is still in its infancy and takes up considerable amounts of energy. It definitely has potential.

Flexible packaging, such as that pet food comes in, or soup, or crisps bags, are often considered impossible to recycle, because they are made of plastic and aluminium laminate. Many companies work on better recyclable alternatives. Additionally, research is going on into alternatives for coloured plastics. Obviously, it is virtually impossible to get rid of mixed pigments in the recycling process. Trials have shown that a coloured coating works. So do shrink sleeves as a replacement for in-mould labels. Self-peeling labels can improve the recycling process as well. These two-layer labels fully stick to the packaging skin at room temperature, but

Mondi has developed a fully-recyclable plastic laminate for pre-made pouches and FFS roll stock that is perfect for integration into existing recycling schemes. BarrierPack Recyclable has been validated for existing industrial recycling streams.

the difference in expansion of the layers at rising temperatures causes the label film to loosen itself to be easily removed.

Used to virgin

Ioniqa is a high-tech chemical company developing, customising and producing their Magnetic Smart Materials & Separation Processes for multiple applications including infinite PET bottle, textile and carpet waste recycling to high grade PET with virgin PET properties. It can compete with real virgin PET in both quality and costs. The technology is capable

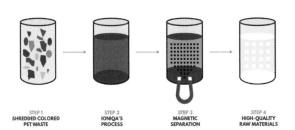

Ioniqa's Magnetic Smart Material Separation Processes

of transforming the PET polyester industry into a circular business.

Laminate material recovery

Suez, a large waste management company, claims to have shown in collaborative research that recycling laminates is possible. They tested various technologies, such as the one by Enval, with microwave-induced pyrolysis. Some hurdles are still to be surmounted, mainly the ones concerning separation of polyolefins from other plastics.

Many more new technologies are awaiting their breakthrough. Some have been described in the previous chapter. It is important to note here that good ideas may be the result of seeing principles in areas that seemed irrelevant. Cross-referencing and open-mindedness are keys to flow improvement. These words point at common ground, an area of cooperation where initiatives can start to find their way.

A man scavenges the streets of Rotterdam, picking up old doors and boards: materials considered worthless by others. He turns them into beautiful and spectacular works of art in his workshop, celebrating the inventiveness and perseverance of humanity as well as its frailty and failures.

HOLOCENE, Ron van der Ende, 1913, Bas-relief in salvaged wood, 160 x 168 x 16 cm.

USA, Utah, Canyonlands National Park, newspaper rock petroglyphs

A photofinish composed of multiple slices of time

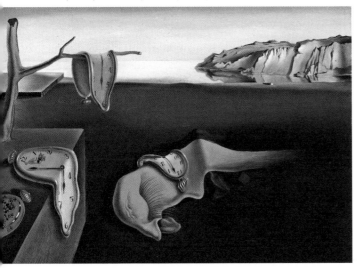

The Persistence of Memory, Salvador Dalí, 1931

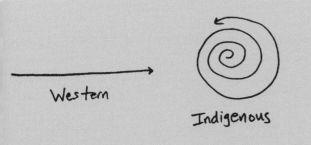

TIME TO RETHINK

Time doesn't flow or move but allowing other things to do so. So cycles that we observe have nothing to do with time other than time allows them to exist. Time is linked to every part of space, forming the space-time continuum. Everything is moving and vibrating, because time allows change, and space allows elements to be in different places. When elements change position time allows that.

Despite our (non-digital) clocks still moving around in circles, our western perception of time is linear. This is what makes us think in terms of progress with an ambitious desire to rule both time and our natural environment by subjecting them to two other conceived economic concepts: property and money.

Indigenous people, in this context an indication of non-westerners, may have a totally different concept of time: everything moves around in a circular or a spiral system of natural and social relations that need to be balanced and protected.

Because of our linear way of thinking the industrial world is facing an ecological crisis. Yet few industrial economists would admit they could learn from indigenous people. Their economies are often called 'primitive', their technology dismissed as 'Stone Age', and most governments assume these people can benefit only from salaried employment. Still, their traditional ways of life have proved highly durable. The key to this success is sustainability. Indigenous people today still use resources without depleting them. They use their intimate knowledge of plants, soils, animals, climate, and seasons, not to exploit nature but to co-exist alongside it. This involves careful management, control of population, they consume of small quantities in a wide diversity of plants and animals, small surpluses, and minimum wastage. They manage 80% of the Earth's biodiversity, yet they only occupy 22% of planet's surface.

Time is one of the most familiar and yet mysterious properties of our universe. The 'flow' of time is one of the strongest impressions we have, yet it may simply be an illusion or a product of the conscious mind. The very notion that time somehow moves leads to a logical paradox because, as the Australian philosopher J. J. C. Smart asked: 'In what units is the rate of time flow to be measured? Seconds per __ ?'

appendix

SOURCES AND FURTHER READING

- Achterberg, Elisa, Jeroen Hinfelaar & Nancy Bocken Master circular business with the value hill, Circle Economy, Amsterdam 2016.
- Adidas Speed factory (website), http://www.adidas.com/us/speedfactory
- Amsterdam University of Applied Sciences, Saxion, Modint, Circle Economy, Sympany and MVO Nederland, Measuring the Dutch Clothing Mountain; data for sustainability-oriented studies and actions in the apparel sector 2017.
- Avery Dennison, RFID Intelligent Labels (website) http://label.averydennison.com/na/en/home/products/intelligent-labels.html
- Bakker, C. Hollander M. den, Hinte, E. van & Zijlstra, Y (2014), Products that Last, product design for circular business models, TU Delft, Delft 2014.
- Bhardway, V. & Fairhust, A. Fast fashion: Response to changes in the fashion industry, The International Review of Retail Distribution and Consumer Research, published online 2010
- Campion, N. et al. Sustainable healthcare and environmental life-cycle impacts of disposable supplies: a focus on disposable custom packs, Journal of Cleaner Production Volume 94, Pages 46-55, 1 May 2015,.
- C&A responsible T-shirt, http://www.mcdonough.com/mbdc-assesses-first-cradle-cradle-certified-gold-t-shirts/ 2017
- CHEP (website), https://www.chep.com/location-gate
- CIRCO, creating business through circular design http://www.circonl.nl 2015-2018
- Coca Cola, (Source: The Guardian, 2017 https://www.theguardian.com/environment/2017/mar/15/millions-of-single-use-plastic-soft-drink-bottles-sold-every-year-report-shows
- Copenhagen Fashion Summit, Pulse of the Fashion Industry 2017
- Cup2Paper (website) www.cup2paper.com/
- Dabbawalas, Mumbai Food Delivery System, (website) Independent, https://www.independent.co.uk/life-style/food-and-drink/dabbawalas-food-delivery-system-mumbai-india-lunchbox-work-lunch-tiffin-dabbas-a7859701.html
- DELL (website), http://www.dell.com/learn/us/en/uscorp1/dell-environment-packaging-and-shipping
- Den Oever, M. et al. Biobased and biodegradable plastics – facts and figures, Wageningen Food & biobased Research, number 1722, 2017
- Ecoeuros, reversed vending machines, http:// www. ecoeuros.nl/hoe-werkt-ecoeuros
- Ellen MacArthur Foundation, Towards the Circular Economy Vol. 1: an economic and business rationale for an accelerated transition 2012
- Ellen MacArthur Foundation, Towards the Circular Economy Vol. 2: opportunities for the consumer goods sector 2013
- Ellen MacArthur Foundation, The New Plastics Economy: Rethinking the future of plastics 2016.
- Ellen MacArthur Foundation, A New Textiles Economy: Redesigning fashion's future 2017.
- Ellen MacArthur Foundation, The New Plastics Economy: Rethinking the future of plastics & Catalysing action 2017.
- EFSA European Food Safety Authority (website), www.efsa.europa.eu 2018
- Eosta, Nature & More, Natural Branding (website), www.natureandmore.com/en/natural-branding
- EU, Circular Economy, Implementation of the Circular Economy Action Plan (website) http://ec.europa.eu/environment/circular-economy/index en.htm 2018
- EU Strategy for Plastics in a Circular Economy (2018), http://ec.europa.eu/environment/circular-economy/pdf/plastics-strategy-brochure.pdf
- Fashion United, Global Fashion Industry Statistics (2018, website), fashionunited.com/global-fashion-industry-statistics
- FilippaK, Second Hand (website), http://www.filippaksecondhand.se/english/
- Freedonia Group, Industry Study, world medical disposables, foodservice disposables and world food containers, http://www.freedoniagroup.com
- FREITAG, the Freitag story https://www.freitag.ch/en/about
- Global Fashion Agenda, Pulse of the fashion industry report, Global Fashion Agenda http://www.globalfashionagenda.com 2018
- GObox, reusable take-away containers (website), http://www.goboxpdx.com
- Heineken FOBO bottle (website), http://www.theheinekencompany.com/sustainability/case-studies/heineken-fobo-the-forwardable-bottle-with-a-story-to-tell
- HIGG Index (website), Sustainable Apparel Coalition, https://apparelcoalition.org/the-higg-index/
- Huffington Post, An American Plate That Is Palatable for Human and Planetary Health, blog by Johan Rockström, Walter Willett, M.D., DrPH, MPH, and Gunhild A. Stordalen 2015

- IBWA (International Bottled Water Association), http://www.packagingtoday.co.uk/features/featureplastic-bottles-getting-stronger-and-lighter-4691165/
- Instock, *rescued food*, http://www.instock.nl/en
- Ioniqa (website), Magnetic Smart Materials and Separation Processes, www.ioniqa.com/
- Katja Gruijters Food Design (website) http://www.katjagruijters.nl/pages/uk/newshome.php
- Keykeg, PET containers (website), http://www.keykeg.com
- Klooster, R. ten (2002), Packaging Design, A methodological development and simulation of the design process, p12, Phd thesis, TUDelft.
- LENA the fashion library (website), www.lena-library.com
- Johnson, Bea, 'Zero Waste Home: The Ultimate Guide to Simplifying Your Life, and 'bulk finder' (website) https://zerowastehome.com/
- Made-by, Environmental benchmark for fibres (website) http://www.made-by.org
- Marks & Spencer, The Guardian www.theguardian.com/environment/2017/jul/18/ms-slashes-plastic-use-in-food-packaging-to-cut-waste
- McDonough, William and Michael Braungart, Cradle to Cradle - Remaking the Way We Make Things, Rodale Press, Emmaus 2002
- McDonough, William and Michael Braungart, The Upcycle, NorthPoint Press, New York 2013
- Monique van Heist, hello*fashion*, (website) http://www.moniquevanheist.com
- MUD Jeans (website), How to lease a jeans, https://mudjeans.eu/how-lease-a-jeans-works/
- National Geographic, What the world eats http://www.nationalgeographic.com/what-the-world-eats
- Niero, M & Hauschild, M. Z. Closing the loop for Packaging: finding a framework to operationalize Circular Economy Strategies, the 24th CIRP Conference on Life Cycle Engineering 2017
- Original Unverpackt (website), http://original-unverpackt.de/supermarkt/
- Original Repack; reusable packaging service for e-commerce http://www.originalrepack.com
- Paperfoam, biobased packaging solutions (website) www.paperfoam.com
- Partners for Innovation, RVO & NRK, Ingeborg Gort & Abel Gerrits (2015), Designing with recycled plastics, guidelines (website) http://www.partnersforinnovation.com/media/Guidelines-designing-with-recycled-plastics.pdf
- Peeze, biodegradable coffee cups, (website) World Food Innovations (website) https://www.worldfoodinnovations.com/innovation/biobased-compostable-single-serve-coffee-capsules
- PBL - Netherlands Environmental Assessment Agency, Rood T and Hanemaaijer A, Opportunities for a circular economy 2017
- PBL - Netherlands Environmental Assessment Agency, Trudy Rood, Hanneke Muilwijk and Henk Westhoek, Food for the Circular Economy, Policy brief 2017
- PlasticsEurope, Plastics: the facts (2017), An analysis of European plastics production, demand and waste data, www.plasticseurope.org
- Raworth, Kate (2017), Doughnut Economics, Seven Ways to Think Like a 21st Century Economist, Random House Business.
- Re:newcell (website), http://renewcell.se/about-us/ 2017
- Rent the runway (website), http://www.renttherunway.com
- Rijksoverheid, Monitor voedselverpilling (in Dutch) 2017
- Robertson, G.L., Food Packaging – principles and practice, CRC Press Taylor & Francis Group, USA 2013
- Seepje (website), https://www.seepje.com/how-seepje-works/
- Splosh, refillable soap and detergents (website), http://www.splosh.com/how-it-works
- Terracycle, Zero Waste Box (website), http://www.zerowasteboxes.terracycle.com
- The Guardian, Suzanne Goldenberg US environment correspondent, Half of all US food produce is thrown away, new research suggests https://www.theguardian.com/environment/2016/jul/13/us-food-waste-ugly-fruit-vegetables-perfect
- Unilever Sustainable Living Plan, Waste & packaging http://www.unilever.com/sustainable-living 2018
- United Nations, Sustainable Development Goals www.un.org/sustainabledevelopment/sustainable-development-goals/ 2015
- Urban Mining Corp (website), MDS (Magnetic Density Separation) technology http://www.umincorp.com/
- WASTED Lab, CITIES Foundation (website) https://wastedlab.nl/nl/
- WEPA Satino Black (website) www.wepa.nl/en/brands/satino-black/7257/take-part-in-the-satino-black-recycling-concept.html

BUY
Make
THRIFT
SWAP
BORROW
USE WHAT YOU HAVE

SarahL.com

THE BUYERARCHY
of NEEDS
(with apologies
to Maslow)

COLOPHON

© 2018 Siem Haffmans, Marjolein van Gelder, Ed van Hinte, Yvo Zijlstra and BIS Publishers.

BIS PUBLISHERS, Building Het Sieraad, Postjesweg 1, 1057 DT Amsterdam, The Netherlands
T +31 (0)20 515 02 30, bis@bispublishers.com, www.bispublishers.com

ISBN/EAN: 978-90-6369-498-2 / NUR-description: Industrial design / Title: Products that Flow
Subtitle: Circular Business Models and Design Strategies for Fast-Moving Consumer Goods

C&A Foundation

The publication PRODUCTS THAT FLOW was supported by the C&A Foundation. C&A Foundation is a corporate foundation, affiliated to global retailer C&A, here to fundamentally transform the apparel industry. As an independently-funded philanthropic organisation, we can focus on longer-term objectives with a healthy appetite for risk. We can form alliances with other brands and retailers in pre-competitive spaces and with actors from across the supply chain to co-create solutions to the industry's biggest challenges. And we can share lessons with others to have a greater, collective impact.

ABOUT THE AUTHORS

This book emerged from an inspiring creative process between: Siem Haffmans - strategic consultant and CIRCO trainer (s.haffmans@partnersforinnovation.com, Marjolein van Gelder - researcher and consultant, Ed van Hinte - writer (ejhint@drs22.nl) and Yvo Zijlstra - graphic designer/image editor (zijlstra@antenna-men.com).

Final text editing: Sandra Rawlin (info@ppsproofreading.co.uk)

IMAGE CREDITS

p4. *GiveBackBox.com Goodwill program,* Amazon - p6. Vintage 1961 Barbie dolls, Mattel - p10. Wheat harvest, John Deer Industrial farming, Container harbour CC; Sao Paulo stock exchance, Wickimedia commons; Production line dairy factory, Shutterstock p12. Chickensoup, HEB; Egg packaging by James Tae; Vintage metal 4 dozen egg delivery box CCO; Doughnut Economics model, Kate Raworth - p14. Foot fashion CCO - p15. Supermarket Jakarta, Shutterstock - p16-17. Childlabour Pakistan, Ilo.org, CCO; Wall street stock market, CCO; Plastic soup, CCO; Cheap fashion, Greenpeace - p20. *Dana,* from the series *Seven days of Garbage* by Gregg Segal - p23. Bioplastic cup, Biopack, Heuft LKX automatic crates inspector for deposit systems - p24. Floppy disks, Flickr CCO - p25. *Junk Funk,* album cover, Riverboat Records; Recycled Orchestra image courtesy of the MIM - p26. Biosunglasses courtesy Studio Crafting Plastics; Paper fashion by Alexandra Zaharova & Ilya Plotnikov; Spray fashion courtesy Fabrican; Selfies courtesy Nordstrome fashion; E-waste necklaces by Marcela Godoy; Tyvek raincoat, Flickr; Shoes, Adidas - p30, Wikimedia CCO - p32. Image courtesy Sun Fruits; Oil painting of oranges by William J. McCloskey - p33. Ben Than Markket, Shutterstock - p34. Smart packaging image courtesy Faller; Image gloves courtesy Glovedepot - p35. Biomass packagings courtesy Leafware, Vegware, Leaf Cups - p36. www.zalando.nl - p37. Baloons, Wikimedia, *Inflatable Flowers,* 1997 by Jeff Koons - p38. Bazaar CCO - p42. Image courtesy Louis Vutton - p43. Images courtesy Hightide Lamps; Dollar Shave Club; Youtube; Atelier La Durance - p46. Image courtesy Autostore Systems; Garment factory in Bangalore, Achyuta Adhvarya - p48. Images courtesy Bodum drinking cups - p49. Image courtesy Monique van Helst - p50. Images courtesy Hewlett Packard; Planting Power - p51. Instock, Algues Atlantica - p52. Fashion Library images courtesy Lena. p53 - Images courtesy Enviropack; Printerest - p54. Two works from the series *Found in Nature,* Barry Rosenthal - p55. Image courtesy Ivago, Ghent - p56. Images courtesy Chep palet and container pooling service - p57 Image courtesy Smeraldinan water - p58. Images courtesy Cup2paper - p59. Images courtesy RET - p60. Street garbage collection vehicle, Philipines, Brian Evans; Wastecollector Amsterdam 1945, CCO; Ragman, Paris, 1899, Wikepedia - p61. Image courtesy Woolworth on-pack recycling label (OPRL) - p62. Images courtesy Teracycle - p63. Images courtesy MUD Jeans - p66. Drone, Shutterstock; Donkey in Fez Morroco, Lucy Patterson; Image bikecourier courtesy Outspoken Delivery cycle logistics Cambridge - p69. Images courtesy Unverpackt - p71. Natural branding, Images courtesy Eosta nature & More - p72. Images courtesy Soap by Mail - p73. Images courtesy RePack - p74. Images courtesy Heineken - p75. Images courtesy Flippa K - p76. Images courtesy KeyKeg - p78. Image courtesy Unilever - p79. Images courtesy Re:newcell - p80. Images courtesy Seepje - p81. Images courtesy Coca Cola Company - p82. Bio-Pods, Katja Gruiters & Fons Broess - p83. Images courtesy Satino Black; Dell Computers - p84. Images courtesy Freitag fashion & shoes - p85. Images courtesy Hipster Health - p86. Images courtesy Puma shoes; Light packaging Images courtesy Bryant Yee, Tea packaging by Tin Chan - p90. Foam bed image courtesy BASF - p91. Ball-and-stick model of molecule, C10H18O5, Wikimedia p 92. Gore-Tex - p93. Images courtesy Misfit Juicery; Materiability - p94. Image courtesy University of Michigan, Cornell University, Arizona State University - p95. CCO; Image courtesyTianjin Yinhual New Material Tech Co. - p96. Shutterstock, Olivier van Herpt p98-99. Images courtesy Wikimedia; Hangerpak - p102. Images courtesy Higg Index; C&A p103. Images courtesy Newton shoes; GreenPack; SPI Sustainable Packaging Industry - p106. Esther Poon, Shutterstock; CCO - p110. Images courtesy Greenpeace; CCO p111. *Online services* image courtesy Uber - p112. Wikimedia - p116-117 Images courtesy Danit Peleg; Adidas; Avery Dennison p118-119. Images courtesy Ioniqa, Mondi - P120-121. *Holocene* [2013] Ron van der Ende; Smithsonian Institute; *Photo-finish of the 100m final in 1948,* Popperfoto/Getty images; *The Persistence of Memory,* 1931 Salvador Dalí - p126-127. Sarah Lazarovic; Urban landscape, CCO.